Road pricing

Theory and practice

Nigel C. Lewis

Thomas Telford, London

Published by Thomas Telford Services Ltd, Thomas Telford House, 1 Heron Quay, London E14 4JD

First published 1993
Second edition 1994

Distributors for Thomas Telford books are
USA: American Society of Civil Engineers, Publications Sales Department, 345 East 47th Street, New York, NY 10017-2398
Japan: Maruzen Co Ltd, Book Department, 3-10 Nihonbashi 2-chome, Chuo-ku, Tokyo 103
Australia: DA Books and Journals, 11 Station Street, Mitcham 2131, Victoria

A catalogue record for this book is available from the British Library

Classification
Availability: Unrestricted
Content: Original analysis
Status: Refereed
User: Traffic engineers, planners and economists

ISBN: 0 7277 2011 2

This book is published on the understanding that the author is solely responsible for the statements made and opinions expressed in it and that its publication does not necessarily imply that such statements and/or opinions are or reflect the views or opinions of the publishers.

Typeset at Thomas Telford Services Ltd, using Corel Ventura 4.2

Printed and bound in Great Britain by Galliard (Printers) Ltd, Norfolk

To my mother and father

Contents

Part 1. Theory

Part 2. Practice

1. Congestion, traffic restraint and demand management

Queues, tailbacks and delays: should we blame accidents, break-downs, diversions, or just sheer traffic volume? How common such occurrences are in just about every urban area during peak periods, and increasingly in off-peak periods. Traffic congestion causes frustration and delay. 'If only the other drivers had stayed at home...'.

1.1. Congestion

Such is road travel during the nineties — that liberation and freedom to travel which progressive technological development has enabled us to achieve. The internal combustion engine combined with civil engineering achievements has afforded people increased mobility to travel in relative comfort within cities and towns and beyond. But increasing affluence, continued growth in economic activity and extended leisure time are generating yet further demands for travel — trips motivated by personal convenience to individuals' desired destinations, and over ever increasing distances.

Increasing population, car ownership and numbers of trips are putting pressure on the road network. Generally, investment in public transport has failed to keep pace and be able to provide an attractive alternative with adequate levels of service and journey times that match individuals' expectations. Consequently there is congestion and pollution in urban areas.

1.2. Traffic restraint

With our ability to increase the supply of physical resources and funding so limited, the only option is some form of restraint — either regulatory traffic restraint or increased management of the traffic demand using market-based pricing mechanisms to achieve better utilization of the existing urban road space.

Where new road links are constructed, some drivers may divert initially to the new roads in peak periods, others may modify the timing of their trips and yet others may switch modes to benefit from the new facility. These actions, described as 'spatial, time and modal convergence–triple convergence' by Downs,[1] create increased use and peak period congestion. This is an example of supply creating its own demand. As congestion then becomes worse, the converse phenomenon occurs: drivers divert to other roads, reschedule their trips or change modes ('divergence').

Many of the less subtle methods of traffic restraint are unpopular, particularly as, for many people, car ownership is a necessity, not an option. In spite of this, traffic restraint is increasingly being attempted in urban areas – mainly through parking policies and charges designed to restrict demand for the available facilities. While all drivers are now familiar with charges, fines and even vehicle removal, in most cases they are not deterred in significant numbers.

1.3. Road user charging

Road and vehicle design and development potentially afford the modern car owner the capability of relatively limitless travel and the freedom of choice of destination. But this freedom is effectively limited by the thousands of others exercising that same freedom. Each vehicle trip has an influence on the pattern of movement of other vehicles.

As traffic flow builds up and congestion occurs, the presence of each vehicle contributes to the slowdown, thereby imposing a delay, and consequently a cost, on other drivers' travel.

Unfortunately, drivers do not perceive the real cost of this. In order to overcome this problem it is important to devise a mechanism whereby each driver clearly perceives this cost and is charged according to the real external cost which he or she imposes (in terms of delay or marginal social cost) on other vehicles. Only when drivers clearly understand these costs and impacts on other travellers will their behaviour in terms of whether they use a given route, at a given time in a given manner, be affected.

1.4. Road pricing systems

A driver's perception of the real cost of travel is key to his or her behaviour. Where costs are not perceived drivers still impose external costs on society in terms of delays and pollution. In order to overcome this, the introduction of a fair and equitable pricing mechanism is essential. This is today's challenge.

Pedestrians and cyclists recognize that unrestricted traffic is troublesome, but when they become drivers they too want unrestricted travel. Free ridership for all is unattainable. The solution we seek has to be via an equitable, differential pricing mechanism whereby all drivers perceive and pay the relative variable economic costs according to their times of travel. As a result, if the revenue from such charges is applied in the user's interest and demand is manipulated accordingly, this will go some way towards improving the shared use of limited transport networks.

The time for urban road pricing has come. A number of experiments are under way, some bolder than others. Technological developments have enabled the testing of many options which previously would have been unacheivable. Such tests show that electronic charging for road use is feasible. Personalized individual transport vehicles in some form or other are here to stay. The challenge for the twenty-first century is to manage or ration the limited supply of road space using appropriate methods of charging through the application of intelligent informatics.

2. Congestion pricing

In many towns and cities congestion is extremely apparent, and for increasingly longer periods of the day. In London, where it is particularly evident, the Confederation of British Industries (CBI) has estimated that it is costing approximately £10 billion a year in terms of delay, lost productivity, extra inventory costs and unproductive use of resources.

2.1. Traffic flow

Traffic congestion is a direct result of increased traffic flow. The volume of traffic on a link consists of a series of vehicles, the drivers of which each want to minimize their own journey time on that road. The speed and flow of the traffic is entirely dependent on the behaviour of each vehicle's driver. Each vehicle's progress, therefore, is necessarily dependent (except on an empty road) on how its driver adapts his or her behaviour to that of other drivers. Thus, congestion is primarily a function of personal behaviour and dynamics.

Where traffic volume is low on, for example, a straight unobstructed road, fast moving vehicles move off in the clear until they find themselves behind slower moving vehicles. These faster drivers close up on the slower ones until they judge that it is safe to pull out and overtake, perceiving clear road space in front of the vehicle being overtaken. This typical behaviour is illustrated by the distance/time trajectories in Fig. 1(a). This is termed free flow.

Under high traffic volumes each vehicle's speed is dictated by that of the vehicle in front. With less free road space there are fewer opportunities for overtaking and slower traffic flows result. With increasing traffic density, drivers' behaviour becomes more critical. Drivers are not sensitive enough to each other's behaviour to drive at exactly the same constant speed as each other. Consequently, one driver's acceleration then deceleration relative to other vehicles requires the following driver to brake, which in turn causes the next driver behind to react, and so on, so establishing a shock wave back through the traffic flow as illustrated in Fig. 1(b). This is termed unstable flow.

As speeds fall, traffic volume falls and density rises. If there is some free road space, some absorption of the shock wave may occur and flows may increase again. However, if the road is near capacity the shock wave may intensify until ultimately the traffic comes to a halt. This then becomes forced flow. Under these conditions it is very difficult to restore free flow conditions. With the formation of a queue, free road space gets taken up and the result is congestion causing increasing delays to each vehicle joining the queue.

Clearly a street network is more complex than a simple road link as it involves not just other vehicles but additional obstructions, hazards (e.g. roadside friction) and road intersections. Invariably, capacity is more directly a function of the intersections than in the case of a link. However, the causes of congestion are similar and also result in the build-up and development of queues.

These three flow regimes (free flow, unstable flow and forced flow) are illustrated in Fig. 2.

In summary this indicates that

- from a situation of steady flow as traffic volume increases, speeds begin to decrease but free flow conditions continue

- with increasing volume and as full capacity is approached, speeds fluctuate wildly and unstable flow conditions occur

- once flow is unstable, both traffic volume and speeds fall and stop-go queues form, reflecting forced flow.

These are similar to patterns which can be described by modern chaos theory. This situation can be illustrated further by inversely plotting the time delay against volume (Fig. 3).

2.2. Congestion economics

In order to determine the economic effects of congestion, it is necessary to determine the value of time to travellers. Clearly, different travellers value their time differently. This value may also vary according to the purpose of each trip. Therefore, we can estimate the external cost of congestion by multiplying the delay by an average value for travel time. If the vehicle operating costs borne by the user and the variable road maintenance cost are included, the overall cost of congestion can be seen to behave as shown in Fig. 4. This figure represents the generalized value known as the average variable cost which increases gradually under free flow, but asymptotically under forced flow conditions.

Thus supply can be represented. If the supply is then contrasted with the demand in the lower part of the curve, the classical supply

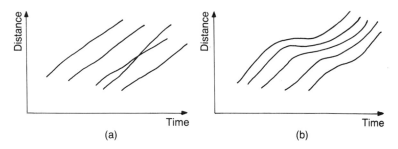

Fig. 1. Distance/time trajectories under (a) free flow and (b) unstable flow showing the shock wave effect

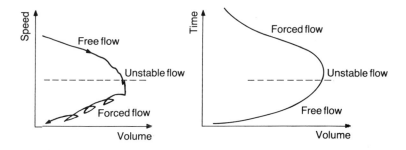

Fig. 2. Speed/volume relationships for the three flow regimes

Fig. 3. Time/volume relationships under three flow regimes

and demand curve is generated (Fig. 5). The demand curve represents the marginal benefit from each additional journey. The marginal cost curves are also included in Fig. 5. These show the private cost of any additional trip as perceived by the traveller, and the higher marginal social cost of the trip. This reflects the cost of delays and external influences on other travellers imposed by this additional trip.

This shows the crux of the whole argument: the difference between the marginal private cost for each journey as perceived by the driver, and the social cost that journey imposes (in terms of delays) on other road users. If this difference is allowed to remain unquantified in the mind of the driver, he or she will still be inclined to make as many

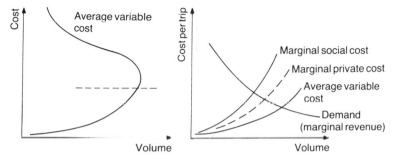

Fig. 4. Average variable cost   Fig. 5. Cost/volume according to volume

trips, thus continuing to contribute to congestion. However, if an appropriate method is introduced in order to charge the road user for this difference, a more balanced level of demand will result. This, then, is the delicate process: to ensure that the mechanism so designed makes drivers actually internalize the external cost of each urban journey.

2.3. Congestion pricing

A driver's decision before undertaking a trip is generally made on the basis of what he or she perceives to be the average variable cost of that trip. This includes the time cost and the vehicle operating cost borne directly by the driver for the trip, and the indirect user contribution made by the driver to road maintenance costs. The resulting external cost imposed on other travellers by the contribution to congestion of the trip is ignored in the decision.

With increasing average variable costs the marginal costs increase at a higher rate. In terms of pricing, the equilibrium point is where the peak demand intersects the marginal cost curve (see Fig. 6). At this point the vertical difference between the marginal cost curve and the average variable cost curve represents the external congestion cost imposed by an additional trip. This therefore reflects the pricing position — that the optimal toll should be based on marginal cost pricing, or the difference between the marginal cost and the average variable cost of a trip, in order to charge for the actual external congestion costs of each trip.

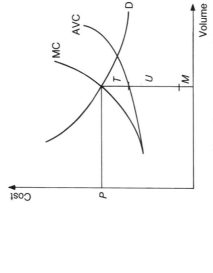

User charge components

T = external congestion cost (toll)
M = road maintenance cost (authority)
U = user operating cost and time (borne directly by user)

User charge = costs imposed on other users and authority
= external congestion cost + road maintenance cost
= $T + M$

Fig. 7. User charge components

ATC Average total cost
AVC Average variable cost
D Demand
MC Marginal cost
RMC Road maintenance cost
UTC User time cost
VOC Vehicle operating cost

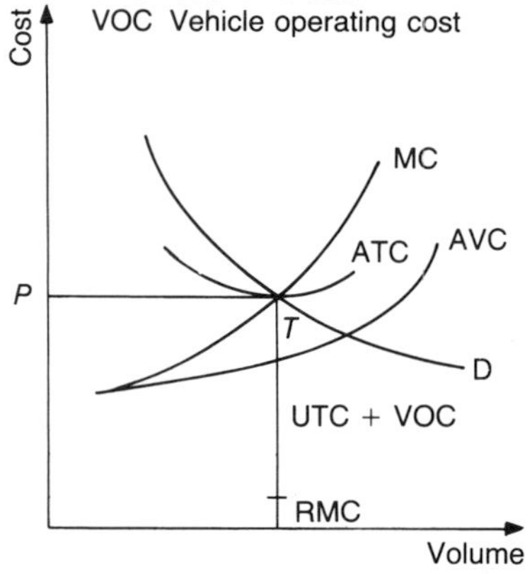

Congestion toll

P = long run marginal cost

Fig. 6. Congestion toll

For each trip the overall road user charge should include components both for the external congestion costs imposed on other users and for the costs imposed on a providing authority. These are illustrated in summary in Fig. 7 along with those other costs (time and direct operating costs) which are borne directly by the individual user.

3. Road user charging and traffic restraint

Most car owners are familiar with the costs of owning and running a vehicle. But from an authority's perspective, there are different ways of charging for the use of a road network. In addition, if desired, traffic restraint can be introduced through a variety of policy measures.

3.1. Road user charges

Methods of road user charging broadly fall into two categories: indirect and direct.

Indirect charges

Generally, indirect methods of charging relate to charges for vehicle ownership and usage.

- Fixed charges for ownership: purchase tax on acquisition and annual licence fees are common in most countries.
- Variable charges for usage: taxes on tyres, spares, oil and fuel are virtually universal, although they are applied at variable and differential rates.

Due to the high volume of usage and the ease of collection, this source (particularly fuel tax) has become relied upon by many governments as a prime source of general revenue. While this tax directly reflects a tax on usage or operation (distance charge), and hence on wear and tear of the road network, in few cases is the revenue hypothecated — that is, channelled directly into a road or transport fund earmarked specifically for highway maintenance or development. Consequently in most countries road users actually contribute through travelling to more than just the marginal physical cost of the general upkeep of the infrastructure.

However, where there is congestion caused by many simultaneous road users, drivers impose a marginal social cost in terms of delay and higher operating costs on other users, for which they are not being charged.

Other variable charges may be applied for usage, such as a daily/hourly parking tax, a daily/monthly area licence charge or a periodic registration charge for a vehicle to be licensed in a particular area. These are effectively location or area charges for usage. The alternative methods of charging are shown in Fig. 8.

In parts of Scandinavia, petrol and diesel are priced differently. Petrol attracts duty, but diesel has less tax applied and consequently is approximately half the cost of petrol per litre. However, diesel powered vehicles are also taxed per kilometre. Sealed units are maintained on each vehicle, which are read each quarter and verified annually, and owners are billed periodically according to their use of the vehicle. This is effectively a fairly sophisticated direct road user charging system (by distance).

Direct charges

Direct charging involves monitoring the actual time or distance of vehicle travel and charging appropriately. With the development of electronic technology, this can now be achieved through the introduction of off-vehicle or in-vehicle systems or some combination of the two, for example a fixed communications beacon with an in-vehicle automatic vehicle identification transponder. These are listed in Table 1.

Table 1. Direct charging mechanisms

Off-vehicle		In-vehicle	
Toll point	Auto-scanner	Meter	Transponder
Manual collection booth	Electronic (gantry antenna	Magnetic card (stored value	Passive Semi-active
Coin operated machine	or inductive loop) Radio frequency	decrements) Smart card	Active
Automatically (dynamically operated)	transceiver for automatic vehicle identification	Clock Odometer	

Toll collection systems may be based on 'point', 'time' or 'distance' collection. Traditional toll collection consists of payment at a point, barrier or cordon for entry on to a particular facility (e.g. a road, bridge or tunnel). More flexible systems consist of payment for time by metering (e.g. taxi systems using a clock). In addition, electronic

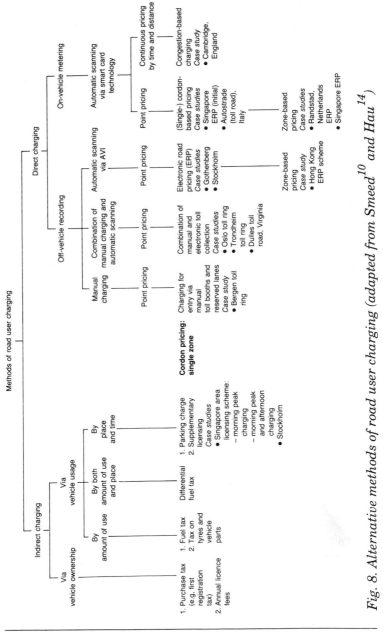

Fig. 8. Alternative methods of road user charging (adapted from Smeed[10] and Hau[14])

systems may charge by distance (e.g. using odometers). Alternatively, some combination of point, time and distance charging may be used. These mechanisms and automatic debiting systems are described in more detail later on.

3.2. Traffic restraint

One of the most effective ways of restraining traffic, and hence minimizing congestion, is through the introduction of a strategy of demand management or appropriate pricing policies for parking vehicles in public roads and streets. Constraints on the supply of parking places effectively restrains drivers from entering urban areas. But on its own this is not adequate because it addresses only the static use of road space.

With increasing growth both in car ownership and car use, municipal authorities have sought direct ways to restrict traffic moving through and into town and city centres. Typical solutions that have been addressed include the physical and regulatory measures given in Table 2.

Downs[1] has summarized the cost-effectiveness ratings of different policy supply and demand tactics that could be applied in addressing a strategy for reducing traffic congestion (Table 3). Some of these are measures which are dependent upon physical planning policy regulation, which may be less effective or more costly.

Many solutions consist of variations of this. Further examples of traffic restraint include measures relating to ownership. Ownership controls have been introduced for example in Bermuda (a limit of one car per household), Japan (a dedicated residential parking space is

Table 2. Traffic restraint measures

Physical	Traffic management and regulatory	Indirect
Construction of bypasses	Area licensing	Improving alternative modes
Road closures and diversions	Restriction of heavy vehicles	Rescheduling activities
Pedestrianization, traffic cells and traffic calming	Bus/taxi only lanes	(flexitime) and teleworking
Out-of-town parking facilities	Park-and-ride	Tax allowances
	Pedestrian/cycle lanes	
	Residential parking and area permits	

Table 3. *Cost effectiveness rating of policies for reducing traffic congestion (adapted from Downs[1])*

Policy*	Effectiveness		Costs		Implementation		
	Extent	Impact	Direct to commuters	To all society	Institution required	Ease of administration	Political acceptability
Supply							
Rapidly removing accidents	Variable	Great	None	Minor	None	Easy	Good
Improving highway maintenance	Broad	Moderate	None	Moderate	None	Moderate	Moderate
Building added HOV lanes	Variable	Moderate	None	Great	Co-operative	Hard	Moderate
Building new roads without HOV lanes	Variable	Moderate	None	Great	Co-operative	Moderate	Poor
Upgrading city streets	Variable	Moderate	None	Moderate	None	Easy	Moderate
Building new off-road transit systems, expanding existing ones	Narrow	Moderate	Minor	Great	Co-operative	Hard	Poor
Increasing public transport usage by improving service, amenities	Narrow	Minor	None	Moderate	None	Hard	Moderate
Co-ordinating signals, CCTV ramp signals, electronic signs, converting streets to one way	Narrow	Minor	None	Minor	None	Moderate	Good

*Listed in categories in descending order of effectiveness.

Table 3 continued

Policy*	Effectiveness		Costs		Implementation		
	Extent	Impact	Direct to commuters	To all society	Institution required	Ease of administration	Political acceptability
Demand							
Instituting peak-hour tolls on main roads	Broad	Great	Great	None	Regional	Moderate	Poor
Parking tax on peak-hour arrivals	Broad	Great	Great	None	Regional	Hard	Poor
Eliminating income tax deductibility of providing free employee parking	Broad	Great	Great	None	Co-operative	Moderate	Poor
Providing income tax deductibility for commuting allowance for all workers	Variable	Great	None	Minor	None	Easy	Poor
Increasing petrol taxes	Broad	Moderate	Great	Moderate	None	Easy	Poor
Encouraging and promoting ride sharing	Narrow	Moderate	None	Minor	Co-operative	Hard	Moderate
Encouraging people to work at home	Broad	Minor	None	None	None	Moderate	Good
Staggering working hours	Variable	Minor	None	None	Co-operative	Moderate	Moderate
Increasing car licence fees	Broad	Minor	Moderate	Minor	None	Easy	Poor

*Listed in categories in descending order of effectiveness.

required for every car), Hong Kong (a first-time registration tax) and Singapore (an additional registration tax and preferential annual registration fee). In Singapore the high cost of ownership is a particularly severe deterrent.

Other measures that act as restraints to usage are fuel tax, differential fuel tax and tyre tax, all of which are used as policy tools to restrain traffic use and are a function of distance travelled.

3.3. Area licensing

Policy methods are emerging that focus on demand and regulatory management of traffic movement, for example

- area restrictions
- time restrictions
- lane-use restrictions
- vehicle class restrictions
- occupancy restrictions
- tidal flow systems
- ramp metering.

All of these are aimed at maximizing the capacity of the available road space by time and place. Further management measures are listed in Table 4.

Municipal authorities are continually wrestling with alternative policies concerning on- and off-street parking pricing schemes. These include zone licences, street parking meters, and vehicle parking discs, tags and stickers. Charging regimes may be varied to include progressive charging rates. Private or public off-street and private non-residential parking are methods which free up the actual public road space.

More esoteric methods have also been proposed, such as the Athens 'even and odd' daily licence plate scheme or the New Zealand trial 'car-less day per week' scheme.

Table 4. Measures to increase road capacity

High occupancy vehicle lanes	Enhanced enforcement of regulations
Co-ordinated traffic signals	
Parking restrictions in streets	Co-ordinated maintenance programmes
Improved real-time information	
	Closed-circuit television monitoring
Restricted street access	

Area licensing schemes have been tried experimentally. These act effectively as admissions charges to central congested areas at peak periods. Generally, daily or monthly licences are pre-purchased at sales outlets and displayed simply as visible stickers on the windscreen which can be verified visually. These schemes are very cheap to implement and are generally effective if there is an adequate level of compliance or enforcement in the area.

Road space is a restricted resource which is competed for by different users. Since there is only limited ability to increase the supply of such space, efforts have focused on improved management of the vehicles using the roads and of peak spreading. While direct user charges may be proposed as an effective charging mechanism for movement on the roads and to suppress no-essential trips, some form of charging for or rationing of parking space is also required. Given that such occupation of the road space restricts its usage by other vehicles, an equivalent charging system by time and class of vehicle is required. For maximum effectiveness both of these elements need to be integrated to form an efficient urban road pricing system.

4. Urban road pricing systems: goals and requirements

In order to overcome traffic congestion, an urban road pricing system can be designed with particular objectives and goals in mind. Its success will be dependent on satisfaction of the defined institutional, financial, operating, technological and social criteria.

4.1. Policy objectives and institutional framework

Problem formulation

While policy is formulated by decision makers at an authority or government level to meet societal objectives, actual response is determined by individual decision makers — the travellers who aim to optimize their travel time and cost for each particular trip. The policy problem is defined in terms of broad objectives. The response problem — reflected in the travel choices of individuals (e.g. route, mode, destination, time) — is solved through individuals aiming to maximize their own utility.

In general, in order to address a defined problem, a successful urban road pricing system must, as far as possible, satisfy the following objectives:

- equity — fair distribution effects on each income/user group
- efficiency in the use of resources
- acceptability to each user group.

These objectives can then be articulated further in various ways.

Objectives and their definition

The urban road pricing system should reflect the particular policy objectives of the authority concerned. This will depend on that authority's own context and its needs as reflected in its transport planning framework. Its generation and distribution of traffic is dependent on topography and land use; residential, commercial and industrial location; existing transport networks; and other generators or attractors of traffic which may be outside its control.

It is important that an authority's transport plan should clearly define its policy objectives with respect to the introduction of a road user charging system. These high level objectives should directly relate to the defined problem. This may directly address the transport sector or it may entail serving one or more economic, environmental or social goal, for example to provide additional revenue, reduce environmental pollution, or reduce congestion.

These goals should be clearly defined and may be achieved through the introduction of selective measures in order to respond to primary transport objectives or other secondary objectives, for example by

- exercising traffic restraint (volume)
- redistributing traffic patterns (by time)
- redistributing traffic flows (by location)
- reallocating trips (by mode).

From a policy perspective, road user charging may be based, for example, on any of the following:

- to act as a rationing mechanism for scarce road space
- to provide a more accurate reflection of the true cost of driving
- to reduce non-essential travel demand and consequently the level of environmental pollution
- to provide a source of supplementary revenue
- to act as a more effective alternative than pure regulation
- to reduce demand for the construction of the new roads
- to generate higher returns on privately financed infrastructure
- to 'level the playing field' between private cars and public transport
- to retain travel freedom of choice.

Institutional framework and jurisdiction

The authority responsible for implementing an urban road pricing system in a given area should have

- overall responsibility for transport policy and finance

- executive responsibility for traffic matters
- general and financial accountability to the level of government responsible for planning and environmental matters.

The fundamental decision framework should be carefully defined for the particular context and goals for which the desired solution is being sought. The authority would then be responsible for

- drawing up proposals and public consultation
- submitting the charging regime for central government approval
- implementing, managing and administering the system
- allocating and earmarking the revenues for adopted transport and environmental improvement programmes.

This would ensure that the benefits obtained from the road pricing scheme revert as directly as possible to the users who are actually charged.

There are certain instruments which policy makers may have available under their jurisdiction in order to address defined objectives. For the particular policy objectives defined, the decision variables should be carefully selected in order to achieve them. The success of the system lies with the authority retaining direct control over its realization.

The adopted goals should be verified by undertaking a cost-benefit analysis of the proposed scheme. In general the benefit of the scheme (savings in travel time and operating costs) should exceed the cost of implementing the scheme. The actual level of pricing should be determined with the view of maximizing the benefit to the population concerned rather than on solely aiming to maximize the potential revenues. In this respect from the authority's, the user's or the operator's perspective, the objectives may be differing or conflicting.

4.2. Financial criteria

Having established the policy objectives and the institutional framework within which the urban road pricing system should operate, the financial parameters that relate to this framework should be decided. Selective charging alternatives are given in Table 5.

The applicable charging mechanism, rates, times and vehicle categories can be varied so as to shape the system to respond to the desired policy framework, whether it is to restrict travel volumes, maximize revenue or redistribute traffic flows. One of the principal targets in aiming to reduce traffic congestion should be to focus on reducing non-essential vehicle trips in peak periods.

User response — that is, the price elasticity of demand for car travel — is one uncertainty that should be recognized. Generally it is found that demand for private car travel is very price inelastic with significant increases in tolls or fuel prices, despite an initial fall, invariably travel reverts to its previous levels over relatively short time periods.

The urban road pricing system to be implemented can be considered in two components: the charging regime determined by the authority and the operating system designed by the supplier.

The charging regime should be formulated towards satisfying the specifically identified policy objectives. It should accord with the general transport policies and programmes adopted by the authority and define the detailed arrangements, methods of application, roads, vehicle categories, jurisdiction, etc.

The following are general principles concerning charging which should be satisfied both from the authority's and the user's points of view.

- Pricing should be closely related to the amount of direct use of the roads (by point, time or distance).

- It should be possible to vary prices responsively for different roads or areas, at different times for different classes of vehicle.

- Prices should be readily ascertainable by users prior to the beginning of a trip.

- It should be possible to pay in advance.

- The incidence of the charging on different road users should be fair.

- The charging system should be user friendly, respect privacy and operate on a reliable basis.

Table 5. Charging alternatives

Objective	Raise revenue	Reduce congestion	Environmental enhancement
Mechanism	Cordon	Area	Area
Charging	Uniform rate	Differential rate	Differential rate/vehicle categorization
Period	All hours	Peak hours	All hours

- The optimal pricing method should be based on marginal cost pricing by setting a congestion charge which is equal to the difference between the marginal cost and the average variable cost of each trip.

4.3. Operating criteria

The operating system should be simple to understand and be robust, with a high degree of reliability.

It should minimize the risk of evasion and have a low operating cost, be compatible with other systems and be potentially developable in a modular fashion to include charging for street parking.

The operating system parameters should be capable of achieving the criteria listed in Table 6.

Table 6. Operating criteria

Reliability	mean time between failures 1 in 1 million
Scale	up to 1000 charging points
Traffic volumes	up to 2000–2500 vph per lane
Traffic speed	up to 160 kph
Traffic data	record classified traffic flow, time and location data (data transmission communication rate 100 kbit/s)
Availability	facility for handling non-equipped vehicles (out-of-town occasionals)
Evasion	violation to be detectable and enforceable and not exceed 1% loss
Transparency	tariff information to be clearly displayed
Privacy	confidential, to record vehicle registration details, not passengers
Flexibility	pre- and post-payment charging methods by time and location
Traffic charges	up to 10 different charge levels and up to 5 different vehicle classes
Simplicity	user friendly, anonymous and unintrusive
Compatibility	standardized and capable of integrating with other systems
Security	secure from theft and fraud using cryptology
Environment	minimum environmental intrusion

4.4. Technological options

The fundamental technical components for an effective automatic debiting system for road user charging comprise

- in-vehicle units to enable communications with roadside (off-vehicle) equipment units

- roadside communication beacons that can communicate two-way (or interrogate one-way) with in-vehicle units under an adopted accounting system

- vehicle presence detectors and enforcement system and

- vehicle-roadside data communication network

In-vehicle units

The type of in-vehicle equipment installed and level of intelligence adopted dictates the functions of an automatic charging system. In-vehicle units may be based (as listed in increasing complexity) on any one of four types

- a read-only system where roadside beacons interrogate an in-vehicle tag for the code identification for a central accounting system

- a read and write system where roadside beacons interrogate a tag and also transmit data one-way to an interfaced 'stored value decrement card', which acts as independent user-held transaction log

- a read and write system incorporating two-way data communication between beacons and an in-vehicle transponder which has an in-built micro-processor, or

- a smart card metering system incorporating two-way data communication between beacons and an in-vehicle transponder, interfaced to a user-held smart card which registers and meters transactions with electronically erasable programmable and read only memory (E^2PROM).

Under the first system, a fixed antenna transmits a radio signal which is modulated by an AVI tag on board a vehicle and reflects its encrypted identification code back to be received by the antenna. This modulated signal is read, decrypted and validated and records the passage of the vehicle on a subscriber's account off-vehicle through the remote computerized recording and accounting system. For this type of electronic toll collection system, high frequency radio transmission, microwave or surface acoustic wave technology may be used.

Generally, the less in-vehicle data processing, the simpler and cheaper the in-vehicle equipment. However, this requires increasing complexity for the off-vehicle and centralized equipment management and processing system. Stored value decrementing cards may be recharged in value at designated sales points. The smart card system, the most complex in-vehicle type, operates as an independent metering system, delegating control over accounting to the user rather than to a central authority. Characteristics of these different systems are summarized in Table 7(a).

Table 7(a). Technical systems and accounting/debiting methods

Technical systems

	Read-only (AVI transponder)	Read/write (stored value card)	Read/write (in-vehicle transponder)	Smart card (meter)
Characteristics	Passive one-way	Limited one-way	Limited two-way	Active two-way
Security	Robust	Sensitive	Sensitive	Sensitive
Reliability	Excellent	Limited life	Good	Limited life
Maintenance	Fit and forget	Limited life	Good	Limited life
Price	Cheap	Reasonable	Expensive	Expensive

Accounting/debiting methods

	Account identification		Subscription		Auto. debit
	Pre.	Post.	Ident.	Anon.	
Voucher	—	—	—	Yes	—
Read-only	Yes	Yes	Yes	—	—
Read/write	Yes	Yes	Yes	Yes	—
Read/write (in-vehicle)	—	—	—	—	Yes
Smart card (meter)	—	—	—	—	Yes

Accounting/debiting options

There are a series of different feasible accounting/debiting options, generally determined on a pre- or post-payment basis, dependent on the in-vehicle equipment installed. These can be designed to be managed by a central authority or to be directly user held, for example

- pre-payment

    ○ central operating authority account (where credit is debited per transaction)
    ○ subscriber (with advance subscription or pass) or
    ○ user-held account (smart card meter with credit debited per transaction).

- post-payment

    ○ central operating authority account (with billing periodically in arrears), or
    ○ bank (direct debit per transaction).

These are summarized in Table 7(a).

Simple read-only in-vehicle tags are available, cheap, proven and reliable. Using this system, processing is done at off-vehicle roadside units and accounting is undertaken by a central authority. Intelligent in-vehicle units that undertake data processing on-board are more complex. Currently, these are mostly in a prototype form and require further development. With a smart card metering system, accounting transactions are undertaken directly on user-held cards which act as individual electronic purses.

Detection and enforcement

For maximum effectiveness, automatic debiting systems dictate particular enforcement requirements. Currently, systems which are available to detect and classify vehicles without stopping are based on inductive loops and axle sensors. These are proven and reliable provided that vehicles are channelled to remain in-lane.

Increasingly the focus is turning to photo-logging and video-imaging of vehicles and the use of automatic licence plate recording to record potential violations of non-payment and to prevent fraud. In this area the technology, accuracy and reliability of alternative systems is developing rapidly.

Vehicle–roadside data communications

The system adopted of vehicle–roadside data communication is dependent on the communications medium used for data transmission. This may be designed using

- passive tags (one-way communication only, no independent power required)

Table 7(b) Characteristics of vehicle – roadside communications[13]

Technological characteristics	Inductive	Infrared	Surface acoustic wave	Radio frequency	Microwave
Communications link Available frequencies	10 kHz–250 kHz	—	400 MHz 815–915 MHz 2.45, 5.8 GHz	915 MHz 400 MHz, 800 MHz	2.45 GHz 5.8 GHz
Range	<2 m	<50 m	<10 m	<20 m	<30 m
Data rate: typical	9.6 kbps	100 kbps	—	10 kbps	100 kbps
maximum	19.2 kbps	500 kbps	—	>10 kbps	>500 kbps
Restrictions on transmitted power	Few	None	Moderate/severe	Moderate	Severe
Reading reliability/accuracy based on a range of manufacturers' figures for single-lane operation	95–98%	Not available	99–99.9%	98–99.99%	98–99.99%
Multi-vehicle/multi-lane communications capabilities (without enforcement system)	Full multi-lane	Problem with multi-signal discrimination	Limited multi-lane demonstrated	Pseudo multi-lane demonstrated	Full multi-lane demonstrated
Susceptibility/immunity to weather	Highly immune	Highly susceptible	Relatively immune	Relatively immune	Relatively immune
Electromagnetic interference	Slight	Severe (sunlight)	Slight/moderate	Slight	Moderate

Table 7(b) continued

Technological characteristics	Inductive	Infrared	Surface acoustic wave	Radio frequency	Microwave
Roadside unit					
Relative cost of beacon unit	Medium	High	Medium/high	Medium	Medium/high
Mounting position	In-roadway	Gantry/post	Gantry/post	Gantry/post	Gantry/post
Relative installation/maintenance costs	High	High	Medium	Medium	Medium
Relative size	Large	Medium	Medium	Medium/small	Medium
Tag/transponder					
Operation*	P, SP, A	A	P	P, SP, A	P, SP, A
Communications potential	Half-duplex	Duplex	Simplex	Half-duplex	Half-duplex
Relative cost	Medium	High	Low	Medium	Medium
Mounting position	Underside of vehicle	Behind windscreen	Behind windscreen	Side window or front windscreen	Side window or front windscreen
Relative size	Large	Medium	Small	Medium/large	Small/medium

*P = passive, SP = semi-passive, A = active.

- semi-passive transponders that either transmit or receive communications or

- active transponders providing simultaneous two-way communication and requiring independent power source.

The communications media may be

- inductive loop (low frequency)

- infra-red transmission

- surface acoustic wave (modulation technique)

- high frequency radio or

- microwave frequency.

The characteristics of these systems and their advantages and limitations are given in Table 7(b).

Recently the European authorities have decided to standardize on the use of a number of frequency bands around 5.8 GHz (microwave frequency) for road traffic informatics communications applications.

Pilot projects

Since the Hong Kong road pricing experiments in 1985 using electronic number plates (see chapter 17), considerable progress has been made internationally in technological developments which are now facilitating the introduction of automated charging systems.

Under the current EC financed Dedicated Road Infrastructure for Vehicle Safety in Europe (DRIVE) programme (initiated by Directorate General 13 of the European Commission), relating to road traffic informatics, there is a considerable amount of research going on. These programmes have been aimed generally at improving road safety, maximizing road transport efficiency and contributing to environmental improvements. Key programmes and pilot projects in the areas of demand management and integrated urban traffic management systems are listed in Table 8. These have addressed automatic debiting systems and electronic tolling systems particularly in urban areas.

In parallel, another initiative PROMETHEUS (Programme for European Traffic with Higher Efficiency and Unprecedented Safety), a research programme initiated by European motor manufacturers in 1986, has been ongoing, working towards the development of the 'intelligent car'. While PROMETHEUS has focused on the intelligent vehicle, the DRIVE programme has been founded on intelligent infrastructure. The most interesting developments that have oc-

Table 8. DRIVE projects

DRIVE 1 (1991)	PAMELA (Pricing and monitoring electronically of automobiles) SMART (Smart cards for travel and transport)
DRIVE 2(1992)	ADEPT (Automatic debiting and electronic payment for transport) GAUDI (Generalized and advanced urban debiting innovation) CASH (Co-ordination of activities for standardization of HADES) ADS (Automatic debiting systems)

curred in Scandinavia, with the introduction of electronic road user charging systems in Norway at Oslo and Trondheim, are described in detail in Part 2. The pilot project in Hong Kong is also described along with the area licensing scheme in Singapore. After many years of successful manual application in Singapore, three demonstration projects with full electronic systems are under test.

Similarly, under the Intelligent Vehicle—Highway Systems (IVHS) programme in the USA, there has been considerable technological progress. In the USA, there has been much interest in advanced traffic management systems, driver information systems and AVI systems, especially on freeways and turnpikes. In urban areas the technology has focused on dynamic collection for electronic toll and traffic management systems at bridge/tunnel crossings.

In Japan the major research programmes have focused on navigation and communication systems, and information systems using roadside beacons. These include

- RACS (road–automobile communication systems)
- AMTICS (advanced mobile traffic information and communication systems)
- VICS (vehicle information and communication systems).

These have been tried at test sites in Tokyo, Yokohama and Osaka.

4.5. Social requirements

A successful urban road pricing system has to address the demands from the perspectives of the different groups that will be affected. Since the tolls collected are effectively transfer payments, it is essential that the money raised is used such that there is an overall net benefit.

The goal is to maximize the total social welfare function. However, it is recognized that costs and benefits to each different group will vary in terms of money and time. Different users have different marginal utilities of time and of money. Users with the highest perceived values of time may be made better off at the expense of others.

The general goal, through transfers and trade-offs, is to ensure that the net gain through any redistribution exceeds the net loss for any particular group (i.e. an improvement in overall social welfare or a positive benefit/cost ratio). The goal of maximizing the benefit/cost ratio will not necessarily correspond with the goal of maximizing the revenue/cost ratio. While an optimal charge is desirable from an economic viewpoint, from a social or practical perspective some deviation from the theoretically economic optimum may be necessary.

The perspectives of different groups can be defined effectively from the viewpoint of a series of overlapping sets of impacted behaviour groups (IBGs). The London Research Programme into road pricing has identified these groups and their respondents broadly as in Table 9.

Table 9. Impacted behaviour groups and respondents

Group	Respondent
Spatial	Land user
Socio-economic	General public and residents
Mobility	Traveller
Journey purpose	Employee/shopper/recreation
Public service provider	Health/education
Transport provider	Operator
Business	Employer

Financial impacts can be determined for respondents in terms of user benefits through reduced travel, time, enhanced journey reliability and reduced vehicle operating cost. Economic impacts on respondents can be determined with respect to area benefits for an authority in terms of improved air quality (reduced emissions), improved noise quality, reduced accidents, and change in land use patterns. From an operator's perspective the goal of maximizing revenue may result in objectives that conflict with those of the authority or the users.

The reaction and preference of each respondent in a group needs to be analysed with respect to

- anticipated attitudes towards the scenario (perceived impact)
- behavioural adaptation towards the scenario (adaptation)
- actual (informed) attitudes towards the scenario (actual impact).

Detailed assessment of the preferences of these individuals and groups in the area in question is required in order to determine the benefits of a system and thus ensure the successful introduction by an authority of an urban road pricing system which addresses the concerns of the majority of the community. Most importantly, an effective public information campaign is required both to consult and inform impacted groups and individuals.

5. Implementation

The real objective in the realization of a successful urban road pricing system is to gain the political acceptance of the system and the willingness of the public to pay the proposed charges. This can only be achieved if people recognize overall net benefits to their own independent interests.

5.1. Political acceptance

The problem has been well formulated as follows by Zettel and Carll.[2] After the introduction of an urban road pricing system people fall into three categories as follows:

- individuals who are 'tolled' are forced to pay for a commodity which used to be free

- individuals who are 'tolled-off' are forced into less desirable modes, routes or times of travel

- those individuals who are 'untolled' or not previously tolled (e.g. public transport travellers) are worse off due to more congested transport.

In general almost every individual, except those holding a very high value of time, is worse off. In order to remedy this, it is essential that the revenues are restored directly to the transport sector through improved roads and environment and improved public transport services.

In establishing priorities, the goal is to maximize the sum of the satisfaction of the individuals affected. This is a behavioural issue. For example, any one person when interviewed may be

- a car owner/car driver/pedestrian/cyclist/bus rider

- an employee/resident/shopper/recreationer

- an employer/public service provider.

In other words, each individual is a member, at different times, of a different class or group of interested people. Where their behaviour is likely to be affected by a particular policy or action, they can be considered to be constituents of different impacted behaviour groups. These may include groups determined by travel mode, occupation, business, etc.

Thus the voices of individuals, once part of such a group, tend to carry more relative weight than each one would individually. Hence an impacted behaviour group is able, through the common cause of its members, to mobilize increasing opinion and influence for or against an action. Through such an approach, these groups are able to lobby and exert considerable force on behalf of their members.

In the transport sector most authorities are familiar with such groups, which may be comprised as shown in Table 10. Any policy initiative therefore has to attract the support of the largest or strongest behaviour groups. This is best achieved through the combination of a series of measures which individually address the requirements of the greatest number or weight of groups, but which also address necessary trade-offs in order to avoid marginalizing any minority groups.

Table 10. Interest groups

Industries	User groups	Environmental groups	Residents
Motor	Motorists	Walkers/cyclists	Pedestrians
Freight	Truckers	Anti-pollution	Recreation
Public transport	Passengers	Rail travellers	Shoppers
Road building	Contractors	Conservationists	Home owners

Alternatively it can be considered that any road pricing system should be broadly 'revenue neutral'. That is achievable where users travelling at peak periods or congested times would be charged more than users travelling off-peak or in uncongested areas.

Recent surveys in the United Kingdom show that there is increasing public support for a package of measures to overcome traffic congestion, which would include road pricing, provided that it was combined with other elements of traffic restraint and enhancement of public transport alternatives.[3]

The most common general requirements cited were

- improvement of public transport alternatives
- restriction of cars in inner city areas
- enforcement of parking controls

- improvement of road links/street intersections and traffic management
- introduction of traffic calming measures
- improvement of pedestrian/cycle routes

To be acceptable, overall personal mobility should not be impeded but clear improvements in congestion and travel conditions should result.

5.2. Distribution of benefits

While the introduction of road user charges serves to make people carefully address their behaviour so as to minimize or avoid charges, once collected the revenues act as an attractive form of additional funds. Understandably, remarkable new interest groups suddenly appear in order to compete for a share of the financial benefits.

Road user charges should be accumulated in a dedicated road or transport fund. Revenues from this source would then be specifically earmarked for expenditure only on agreed and mandated investments. This linkage would contribute to the acceptability of the scheme.

As an approach for the admission of expenditure, the most eloquent proposal to date is Goodwin's 'rule of three' (see Fig. 9) whereby the benefits from road pricing are distributed through the re-allocation of the road space released and allocation of the revenue earned.[4]

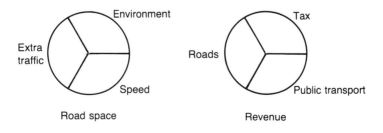

Fig. 9. The 'rule of three' (after Goodwin[4])

Under this proposal reallocated road space is planned to be used for

- environmental improvements (e.g. for pedestrians) and/or
- provisions for extra traffic attracted by speed reductions but not deterred by additional cost (e.g. for high occupancy vehicle users) and/or
- provisions for managing extra speed (e.g. for traffic management systems to handle tendency for traffic growth).

It is proposed that part of the revenue should be spent on

- general local requirements (determined according to local spending plans and priorities) and/or
- new road infrastructure (in accordance with local priorities), and/or
- improvement of the effectiveness of public transport (through service level improvements).

This can serve as a viable starting framework for determining the appropriate distribution of the benefits from the revenue. It should be translated into a series of measures which directly responds to the aspirations of individuals and different impacted behaviour groups, in order to gain their support.

Quite clearly, anomalies in any cordon- or area-based charging system are likely to occur. The main aim is to introduce as equitable a system as possible by taking a pragmatic and flexible approach in its implementation.

6. Summary of Part 1

In order to be successful the introduction of an urban road pricing system should address the issues as set out below as a series of checklists

Policy framework

- The system should be part of an authority's integrated policy framework of a series of measures for an area
- Each measure in the package should provide a net benefit to the community resident in the area
- The overall package should maximize the social welfare of the community within the authority's jurisdiction
- Each measure in the package should be realizable within the plan's time horizon
- Road space development priorities should be focused generally on
  - reducing congestion and improving journey times
  - accommodating traffic growth by improved traffic management
  - environmental improvements and traffic calming.

Charging regime

The urban road pricing system should

- relate to vehicle use of the road space (utilization and location)
- include an associated parking charging system on streets
- be managed and controlled locally under the authority's supervision
- be fair and enforceable
- be based on differential charges for different vehicle categories at different times of day

- be based on differential charges for different vehicle categories at different times of day
- incorporate optimal pricing based on setting a congestion charge which equates to the difference between the marginal cost and average variable cost of each trip.

Operating system

- The operating system should be robust and reliable, simple and flexible
- The operating system should respect privacy
- Information on the charging rate being levied at any time should be clearly outlined to travellers prior to the start of a trip
- The charging system should accept pre- or post-use methods of payment.

Revenue expenditure

- The economic benefits (net surplus revenue from user charges for the system) should be spent directly on
    - the transport sector within the area and/or
    - public transport services, road improvements and traffic management schemes, and/or
    - considered as additional local revenue available for expenditure in accordance with the authority's defined priorities.

7. Oslo

7.1. Introduction

Greater Oslo, the capital of Norway, is a city of approximately 700 000 inhabitants with vehicle ownership of approximately 290 per 1000 people. It is physically characterized by delightful but difficult topography around the Oslo Fjords (Fig. 10). The total volume of traffic towards the city on a typical weekday in 1990 was 230 000 vpd, with highest flows on the western access road of 32 000 vpd, or peak hour volumes of 3000 vph.

7.2. Objectives

In order to improve deteriorating traffic conditions, a new 3 km tunnel was planned to be built under the central part of Oslo, where the topography physically restricts any alternative widening of the road arteries. Due to the high cost of tunnelling and the need to ensure acceptable environmental conditions, it was proposed to raise additional funds to build this tunnel section by toll collection.

Traditionally, government-allocated funds have focused on road development and enhancing accessibility in rural areas. In this case, however, the goal was to undertake a major medium-term development programme of a series of 50 road construction projects (including 13 tunnels) valued at NOK 10 billion (1987), to be financed partly by the government (45%) and partly by road users (55%). Subject to parliamentary approval it was proposed that toll financing (managed by the City Council of Oslo and the neighbouring County Council of Akershus) should be introduced in order to

- finance major road projects in Oslo and the surrounding County of Akershus within an established master plan
- establish improved connections between the fjord and the city
- enhance general traffic flows while giving conditions for public transport and pedestrians the highest priority.

Fig. 10. Oslo's approach roads showing toll points (numbered)

7.3. Institutional framework

A white paper was presented to the Norwegian parliament (Storting) in 1986 which proposed a programme of road developments in Oslo financed through toll revenues. After political approval by Oslo City Council and Akershus County Council, the Storting ratified the scheme. In 1988 the local councils proceeded with plans for the acquisition of land for the toll plazas and a staged development plan was prepared for the overall programme, taking into account local and regional (Scan-link) requirements.

The Ministry of Transport and Communications established the general principles under the Public Roads Act[6] for the toll funding system. These are summarized in Table 11.

Their application is monitored by the Directorate of Public Roads (Vegdirektorat) through the National Public Roads Administration.

A toll system operating company was established through an agreement with the Directorate of Public Roads. This company was to be primarily responsible for administering the collection of tolls and managing the toll revenues, but also effectively to be a financing company for the development of the main road system. Any future toll rate adjustment had to be approved by the County Roads Administration and the Directorate of Public Roads.

The company, A/S Fjellinjen, is a limited company owned by the two main authorities (66% by Oslo City Council and 44% by Akershus County Council) and is wholly responsible for the management of the toll collection system. Under its statutes it is required to publish annual reports and has a mandate to operate for 15 years. The County Public Roads Administration remains responsible for the eventual building of any new road sections, as funds are made available through the toll revenues.

7.4. Financial framework

The aim of the development programme is to improve the highway system. In addition to the NOK 10 billion planned cost over 10–15 years (45% from government and 55% from tolls), 20% of the revenue will be spent on enhancements for public transport (e.g. bus lanes).

Table 11. Toll system principles

Under the Public Roads Act[6] A/S Fjellinjen is authorized to set tolls and allocate funds raised to improve national roads. It is *not* author-ized to contribute to public transport expenditure, or to regulate traffic.

The principles outlined under the Act are

Roads to be funded	Road users to benefit directly from new road projects.
Maintenance	Road building to be in accordance with a master plan.
Time graded tolls	Time graded tolls to be set at local level.
Conditions	Toll company to be responsible to control costs.
Length of toll period	15 years.
Responsibility	To be able to extend toll period if revenue is decreasing. Toll rate or toll period to be reduced if revenue is increasing.
Winding up	When project is paid off
Number of toll stations	Certain distance to be guaranteed between toll stations.
Collection methods	Toll company to be responsible for cost control.
Funding	Tolls to cover at least 50% of investment costs.
Loan guarantee	State loan guarantees. No foreign loans
Levy parallel to construction	Collection of advance tolls in parallel with construction.
Organization	Toll company to be wholly owned by municipalities. Company to be a limited liability company.
Return	No dividend to be paid out.
Tasks and responsibilities	Obligation to generate funds at least cost, to collect tolls and to administer toll revenue.
Toll rates and discounts	Only two toll rate groups (light and heavy). No tolls on people.
Exemptions	Exemptions for public service and emergency vehicles.
Non-payment	Surcharges for non-payment of tolls.
Toll rate levels	Rate not to exceed savings by road user
Authority	Ministry to approve rates as recommended by Directorate of Public Roads.

Table 12. Tariff structure, Oslo

	Light vehicle NOK	Heavy vehicle NOK
Per pass	11	22
Per month	250	500
Per 6 months	1350	2700
Per 12 months	2500	5000
Per 25 trips	230	460
Per 175 trips	1450	2900
Per 350 trips	2600	5200

In order to collect the tolls an entire cordon was placed around the city and a charge made for every vehicle entering the cordon. Construction of the toll plazas cost NOK 110 million and equipment cost NOK 140 million. Annual operating and maintenance costs are NOK 66 million or a cost per transaction of approximately NOK 1.3. Total annual average daily traffic is 210 000 vpd.

Annual revenues being collected currently average approximately NOK 600 million. Consequently, a considerable surplus is currently being earned. Since opening in 1990, tariffs have been increased once by 10%. Currently tariffs are set at a uniform fixed rate for two types of vehicle. Travellers may choose how to pay from the options shown in Table 12.

Tariffs are levied 24 hours per day, every day of the year for any vehicle entering the city. There is no toll for traffic going out of the city centre. Public service vehicles, emergency vehicles, vehicles with disabled drivers, motorcycles and cycles are not charged for. A fixed period card enables an unlimited number of crossings during the validity of the pass. This is attractive to the user as it affords, in effect, a zero marginal cost for the frequent traveller for each additional trip. Average tariffs are given in Table 13. These effectively diminish under the longer calendar time scheme. A case has been made for charging a higher toll at peak periods, as it has been estimated that a toll of NOK 25 would be required to approximate to the marginal external cost.

7.5. Technology

The adopted solution (using Q-FREE by Micro Design AS) was to surround the city with a cordon (toll ring) and install toll collection points at each road crossing for all vehicles entering the city. There are currently 19 stations of 2 or 3 lanes, and 5 or 6 lanes depending

Table 13. Average tariffs, Oslo

Average tariff/trip	Light vehicle NOK	Heavy vehicle NOK
Cash	11.0	22.00
Per 25 trips	9.20	18.40
Per 175 trips	7.71	15.47
Per 350 trips	7.43	14.85
Per 1 month(30 trips)	8.33	16.66
Per 6 months	7.50	15.00
Per 12 months	6.94	13.88

on the traffic volume. This section details some of the system's components. At each station there is a combination of automatic, coin and manual collection booths.

Toll station

These comprise a manual booth staffed 24 hours per day, a coin booth with a facility into which the correct change must be dropped, and an automatic lane (Fig. 11). The automatic lane is part of a dynamic payment system for vehicles installed with a transponder which acts as an AVI system. There are no barriers, only traffic signals.

Registration

The dynamic collection system includes the following.

- In-vehicle equipment for subscribers consisting of a tag on the windscreen which contains a passive transponder. This simply reflects the vehicle's unique identification number.

- Fixed equipment on each automatic lane gantry includes an antenna aerial linked to a registration unit via a transceiver. When a signal is reflected from a passing vehicle's tag, this registers the pass on the toll plaza computer. This is linked via data communication channels to the central computer's management system which automatically debits the subscriber's account. At the toll plaza in the automatic lane a rapid response light signal is given to the subscriber in passing, indicating the status of his or her account.

In addition, any illegal passing through any of the toll lanes activates the video camera which records the licence plate number of the vehicle, the time and location. The system as a whole is illustrated in Fig. 12.

Management system

The management system manages the financial accounting of the system for subscribers. Electronic monthly billing is cost effective. The system caters for the different charging methods which include

- users with a given number of pre-paid trips
- users with a given time period of an unlimited number of prepaid trips and
- invoicing a number of trips over a given time period (post-payment).

7.6. Social implications

Before charging was introduced

At the outset, road user organizations generally considered that an additional road user charge would not be fair, because road users in Oslo already contribute approximately NOK 2.5 billion in taxes whereas only NOK 300 million is actually spent on roads. The cost of collection was also considered to be high (NOK 66 million p.a.). However, as a result of an intensive marketing campaign, road users were lobbied and drivers were offered a 20% discount when signing a subscriber's deal before the opening of the system. This proved effective in helping to gain acceptance of the system.

After charging was introduced

After the opening of the system, approximately one-third of those interviewed were found to be in favour of the toll ring and two-thirds not in favour. Approximately 50–60% of traffic now passes through the subscribers' automatic lanes causing the least delay and inconvenience, and this is increasing.

In a 'before and after' survey of changes in the population's travel behaviour, the following results were noted:

- Mode of transportation — no significant changes
- Travel frequency — general reduction in number of car trips per day from 1.49 to 1.31 and a reduction in overall trips per

(a)

(b)

(c)

Fig. 11. (a) Three-lane — automatic, coin and manual; (b) two-lane — automatic and manual; and (c) coin collection lane

day from 2.66 to 2.37 (some impact also due to recession and increased fuel cost)

- Travel pattern — no noticeable change
- Travel purpose — some evidence for change regarding travel for recreation but not for commuting or in the rush hour
- Impacts on different social groups — high status occupational groups were least affected. Some reduction in road use by younger people evident; some tendency towards out of town shopping and for recreational activities and a tendency to stay within the cordon.

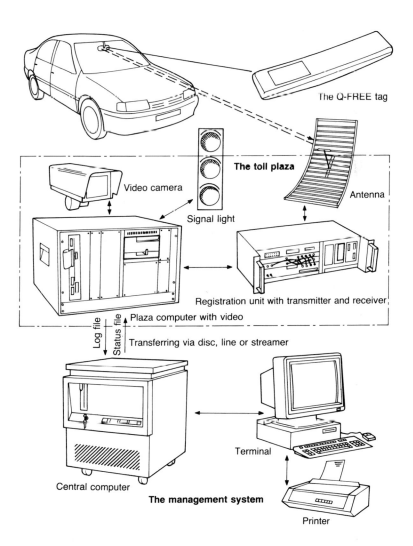

The Q-FREE tag

The toll plaza

Video camera

Antenna

Signal light

Registration unit with transmitter and receiver

Plaza computer with video

Log file

Status file

Transferring via disc, line or streamer

Central computer

Terminal

The management system

Printer

Fig. 12. Oslo's automatic debiting system

The toll ring was opened in February 1990 without the electronic payment system. Consequently the system was initially very dependent on video checking in order to stop violations. Electronic tags were introduced in December 1990. In January 1991 prepaid subscriptions were increased by 14%. In October 1991 an electronic clipcard was introduced — a multi-trip pre-payment system which proved to be popular.

7.7. Conclusions

With respect to the stated policy objectives, the Oslo toll ring system is achieving its objectives of raising additional revenues for a limited period for the development of the main road system. As a policy issue it is likely that pressure may result for more expenditure to be diverted towards public transport rather than to road improvements alone.

In terms of equity, the distributional effects have not been significant. For travel pattern, purpose and frequency, there has been more effect on those in lower status occupational groups and on younger people, particularly for recreational car trips. There has been no noticeable change of mode or increase in use of public transport by any group.

With increasing automation and numbers of passholders, the efficiency of toll collection is improving. The initial equipment installation costs were high. The automated system now affords greatly improved flexibility should a variable toll system need to be implemented. This would more appropriately address peak period congestion costs. To date, the system is not apparently acting as a real restraint to traffic.

Approximately two-thirds of the inhabitants are not in favour of the system. The technology is able, if required, to facilitate higher charging at peak periods to act as a restraint and, as a trade-off, some toll-free or lower charges at off-peak periods. In the medium term, such a policy will probably be desirable if the system is to be acceptable.

One of the difficulties of using a fixed-period discount card is that, while useful in gaining public acceptance, it does not enable the full economic cost of a trip to be charged. This results in regular discount subscribers who travel at peak periods paying effectively less per trip than off-peak travellers — contrary to the requirement of a system for combatting congestion. Discounted trips are better limited to off-peak periods when spare road capacity exists. This would be beneficial for alleviating traffic congestion.

8. Bergen

8.1. Introduction

Bergen is a city of approximately 220 000 inhabitants with a further 110 000 in the Greater Bergen area. Vehicle ownership is around 320 cars per 1000 people. It is a historic city, with its centre projecting into the Byfjorden (Fig. 13). Due to its location it has become a centre of coastal trade. With its topography, nestling by the fjord between the surrounding hills and mountains, access to the central historic area is effectively determined by four roads leading into and out of the city via two bridges to the south and the E14 road to the north. The total volume of traffic entering via these routes is 65 000 vpd. Of this, two locations (to the north-west and south-east) carry approximately 25 000 vpd.

8.2. Objectives

In order to improve access and traffic flow around Bergen, a new tunnel bypass, the Floyfjells Tunnel, was planned. It was constructed and opened in 1986, and provides considerable relief for through traffic. To finance this and other main roads to and from the city, it was proposed to introduce tolls on all such routes leading into the city. In addition a restrictive parking control system was introduced and charges were levied for parking in the majority of the central areas. With this tunnel now complete, the toll system is to be maintained for a further 15 years in order to finance this and other road improvements. The planned development programme is based on the transport plan for Bergen which has identified a series of priorities for individual road sections, tunnels and bridges to be built in the area of the Bergen Commune up to the year 2000. The bridges will have their separate toll fees which will correspond to the former ferry charge.

8.3. Institutional framework

In 1983, a master plan for the development of roads in the Bergen area was prepared by the County Roads Administration. This outlined projects to a value of NOK 2 billion and constituted input to the (four-year) national roads plans.

The toll system was initiated in January 1986 with parliamentary approval having been granted in June 1985 to implement the scheme as proposed by Bergen City Council. As in Oslo, the system is regulated by the Directorate of Public Roads. The toll system operating company responsible for collecting the tolls is Bro-og Tunnelselskapet A/S. It was already in existence (originally established in 1953) and had previous experience in toll collection with the Løvstakk tunnel at the south-western approach to the city which was incorporated into the toll ring. It is a limited liability company, owned 51% by the Municipality of Bergen in the County of Hordaland. For any expenditure planned for new road construction as part of the development plan for Bergen, the government agreed to provide a matching grant for each krone raised through road user charges (tolls).

At the time of installing the toll system a major tunnel was being completed. Since additional funds were available to be applied to the completion of this project using the toll revenue, this, in the public's perception, showed tangible benefits from the new system of tolls. This assisted in gaining acceptance of the toll system. It is unlikely that introducing the toll system initially as a traffic restraint measure would have gained the necessary degree of public support.

Traffic restraint was addressed through an active parking control policy implemented by zone by the Commune of Bergen. Parking is permitted generally for residents only and visitors' parking is restricted to areas with meters or ticket machines in selective zones.

In general, congestion relief and accessibility in the central area has been, and will be, considerably enhanced by the construction of new tunnels: Løvstakk (opened 1986, 2000 m long), Floyfjell (opened 1988, 3800 m long), Damsgardsfjell (opened 1992, 2300 m long) and Nygårdshoyde (opening 1995, 800 m).

8.4. Financial framework

The toll system operates from Monday to Friday between 0600 and 2200 hours for all traffic entering the city. There is no charge on leaving the city. On Saturdays, Sundays and public holidays there is no charge and for public service vehicles there is no charge at any time.

The tariff is a uniform fixed rate per vehicle: NOK 5 for light vehicles and motor cycles (50 cc) and NOK 10 for heavy vehicles (3500 kg). Tolls are paid in cash or by ticket at the toll booth and there is a 10% reduction for pre-purchasing a book of 20 tickets (available at the toll booth). Alternatively monthly, 6 monthly and 12 monthly

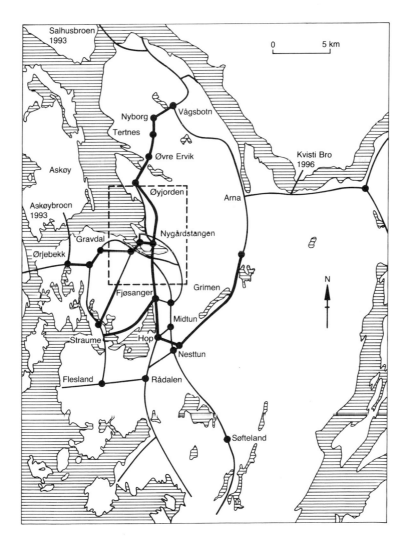

Fig. 13. Bergen's major access road links (inset detail in Fig. 14)

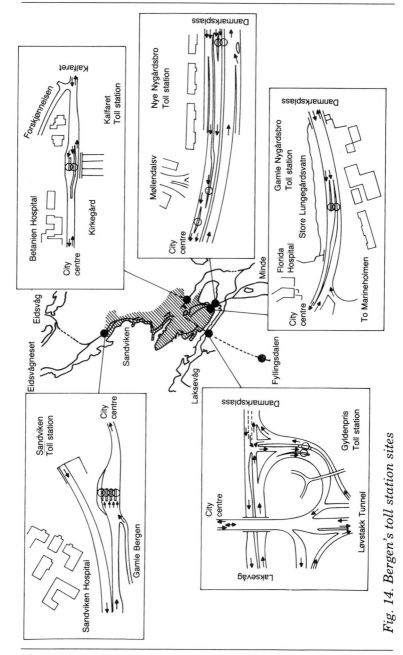

Fig. 14. Bergen's toll station sites

Table 14. Tariff structure, Bergen

	Light vehicle NOK	Heavy vehicle NOK
Per pass	5	10
Per 20 passes	90	180
Per month	100	200
Per 6 months	575	1150
Per 12 months	1100	2200

passes can be purchased. These are displayed on the vehicle wind-screen. The full tariff regime is given in Table 14.

The overall revenue and costs of the system are

- annual revenue approximately NOK 60 million
- operating costs approximately NOK 10 million
- net income approximately NOK 50 million.

The toll rate is set too low to be a congestion alleviating measure. On economic grounds it could be argued that higher rates should be charged in peak hours with longer off-peak periods of non-payment, but the law only allows for toll charging as a road financing measure, not for traffic regulation. It is anticipated that the tariff may be inceased in the future, but this would require central government approval. Operating costs are low at NOK 0.5 per transaction — half the cost of the Oslo or Trondheim schemes (see Table 15).

Revenue raised will be spent on the proposed road development programme over the following four years. Options under study

Table 15. Average tariffs, Bergen

Average tariff/trip	Light vehicle NOK	Heavy vehicle NOK
Cash	5.00	10.00
Per 20 passes	4.50	9.00
Per month	5.00	10.00
Per 6 months	4.79	9.60
Per 12 months	4.58	9.16

and electric buses (trolleybuses in the centre, diesel in the suburbs) if expenditure can be broadened to include public transport.

8.5. Technology

The toll collection system is very simple. There are six toll collection points located on the periphery of the city: one to the north (Gamle Bergen) on the main E15 road; one to the east (Kalfaret) also on the E15; one to the south on the Puddefjordsbroen (555) bridge; one on the Nygardsbroen (556) bridge and one at the Løvstakk Tunnel (see Fig. 14).

Toll collection is entirely manual, via cash payment or prepaid ticket or monthly pass (Fig. 15). Drivers with passes can use reserved lanes and drive through the stations without stopping. The objective of using this system was to increase the use of passes in order to minimize the extent of manning at the toll booths and maximize the throughput capacity. Currently, the traffic consists of approximately 60% pass holders. At peak periods pass holders constitute nearly 80%. Toll collection points comprise either four lanes (two for pass holders, two for those paying cash) or two lanes (one for pass holders, one for those paying cash) with the cash payment lanes on

Fig. 15. Quadruple manned lanes, Bergen

the outside — the inside lane is reserved for buses. Unmanned lanes are equipped with video cameras in order to compare licence plates with registered pass holders. Under Norwegian privacy law, recording can only be undertaken for 5% of the time at any one location. Non-paying motorists, if they are found to be violators, are liable to a fine of NOK 200. This system operates as for parking fines. Income lost through the system is considered to be approximately 1.2% of the total.

Initial congestion problems at the toll booths, resulting from drivers renewing monthly passes on the first of each month, have been overcome by introducing facilities for pre-purchasing passes via banks.

8.6. Social implications

The scheme was introduced after a planning period of just six months in 1985. An intensive public information campaign was set up. Initially reaction was negative. However the toll tariffs were introduced at a relatively low rate with long periods of exemption (i.e. all weekend and at night). This was necessary in order to gain the public's acceptance. In addition, as stated, linkage with completion of a new tunnel led to early public acceptance of the potential benefits.

Since 1986, tolls have been maintained at that same rate and consequently are not perceived as punitive. The current issue is how much can the tariff be increased overall and at the peak in particular, and whether expenditure can be broadened to cover public transport as well as road improvements.

Topographically the cordon is very tight and there are no alternative routes for entering the centre of town. One issue relates to jurisdiction, where the central area tolls are generating revenue for road improvements which are located outside the central business district boundary. While this would seem appropriate for bypasses or environmental improvements, the benefits should be seen to be applied to the area from which the income was derived.

Traffic in and around Bergen is dependent on having a functional regional road network. In the surrounding areas major new bridges are being built (e.g. Askøybrua — which has its own toll regime of NOK 100). Under the proposed development plan the percentage of tunnels on the trunk roads (180 km) will rise from 5% to 25%. The construction of these new tunnels is perceived by the public to improve accessibility and reduce journey times, air pollution and traffic noise, as well as other benefits.

In terms of land use, there is a tendency for activity to move outside the central area. There is some evidence that the toll system may be contributing to acceleration of this process in the development of out-of-town shopping and at regional centres.

As a result of the introduction of the toll ring it was anticipated that motorists might change their route, mode of travel, trip frequency or timing. In a before-and-after survey of drivers in Bergen, the toll ring appeared to have the following impacts:

- there was no evidence of impacts on public transport
- there has been some shift in the timing of trips
- there has been some reduction in the number of car trips, mainly by drivers paying per trip rather than by those with a seasonal pass (for which the marginal cost per trip is zero)
- there are no alternative routes or opportunities for diversion.

8.7. Conclusions

The Bergen toll system has achieved the stated policy objectives of raising additional finance for further road building. The revenue is enhanced by the allocation of a matching government grant for each krone raised via road user charges.

There have been no significant distributional effects since there are no alternative routes for diversion. There is no restraint on shopping or recreational trips at weekends. In terms of equity, the tolls are neutral and are applied similarly to all types of vehicle.

The adopted toll system is not necessarily an efficient means of raising finance. However, the toll system is very simple and does help to minimize the cost of collection, given the revenue objectives.

With respect to the marginal cost of public funds, the cost of tax financing may exceed the cost of toll financing. The economic solution may be a balance of financing partly by taxes and partly by tolls. This requires a more detailed economic analysis.

Currently, over half of Bergen's residents are in favour of the system. The simplicity and equality of the system enhances its acceptability to the public.

9. Trondheim

9.1. Introduction

Trondheim is a historic centre located on the coast of middle Norway. The triangular central area is bordered by the River Nidelva as it turns in its flow into Trondheimsfjord. The city has a population of only about 140 000 (the neighbouring Sør Trøndelag county has approximately 240 000). Like many Norwegian cities, the transport and access network is strongly determined by the topography, with a radial nature to most roads leading to and from the centre (Fig. 16).

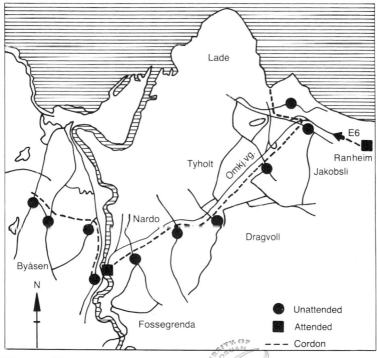

Fig. 16. Trondheim's toll ring

The inhabitants have become accustomed to tolls in the form of ferry fares, and on the adjoining E6 main road between neighbouring Ranheim and Stjørdal. This toll was introduced in 1988 to finance improvements to that road.

9.2. Objectives

As a comparatively small city with no surrounding main road system, the city decided to accelerate the development of road improvements through generating its own funds via road user charges in order to overcome gradually deteriorating traffic and environmental conditions and inadequate road capacity. Its objective in implementing a cordon ring toll collection system was to raise funds for a package of measures that included

- bypasses and relief roads
- cycle and pedestrian routes
- bus-only lanes
- environmental improvement and accident avoidance programmes.

The target was strictly to improve the quality of the transport system and the environment to provide traffic relief in the city. A NOK 2.2 billion development programme was outlined which was to be financed by the toll charges plus a supplemental government contribution. The particular goals were that the toll system should have low operating costs and a non-stop electronic payment system which would maximize road capacity utilization — thus enabling higher peak hour charging — and that it should develop into an integrated transport payment system.

9.3. Institutional framework

In 1985 the City Council developed a transport plan for Trondheim. The Public Roads Administration follows a four-year planning period for expenditure. In January 1987 the City Council decided to introduce a local road user charge in order to address development expenditure, provided that the government would provide matching funds for any future expenditure.

Given the experience of the Trøndelag Bomveiselskap A/S (the toll road company) in operating the E6 facility on the Ranheim to Stjørdal road, this company was given the responsibility of introducing and managing the wider toll ring collection system. However, it was necessary to extend the scope and duration of its mandate. As for

Oslo and Bergen, the operating statutes of the toll company are subject to the regulation of the National Public Roads Administration, which has to approve any changes in tariffs.

The ownership of the toll road company is as follows

- 60% Trondheim City and Sør Trøndelag County
- 40% industrial, trade and truck owners associations and other individuals.

The toll cordon was opened in October 1991. Since the company has become responsible for two facilities (E6 and the toll ring), representation on the board has changed with increasing emphasis on city issues. However, there is some difference between the objectives of the two facilities. Currently, tariffs on the E6 facility are not adequate to cover repayments of the debt already incurred despite being double those of the toll ring. Tariffs are NOK 20 and NOK40 for light and heavy vehicles.

The toll company acts as both a toll collection company and financing company for future developments. Understandably, political pressures have developed to influence the allocation of future expenditure. It was agreed initially that 20% of toll revenue should be spent specifically on public transport, environmental and safety measures.

9.4. Financial framework

The goal of the toll collection company was to make toll collection as automatic and as efficient as possible. The principles of the system are as follows:

- Operation on weekdays only 0600 – 1700 hours (Monday to Friday) for all vehicles entering the city.
- Differential time charges from 0600 – 1000 and 1000 – 1700 hours.
- No charge at weekends, on public holidays or during evening/night periods.
- No charge for public service vehicles or motor cycles.

The tariffs are as indicated in Table 16.

Regular subscribers are charged for 75 entries per month through the cordon — beyond this number, entries are free. This number of 75 (raised from 50) was selected effectively to ensure that the marginal cost of each trip is perceived by individual drivers.

Table 16. Tariff structure, Trondheim

	0600–1000 h Light vehicles NOK	Heavy vehicles NOK	1000–1700 h Light vehicles NOK	Heavy vehicles NOK
Cash	10	20	10	20
Pre-pay 500 passes	8	16	6	12
Pre-pay 2500 passes	7	14	5	10
Pre-pay 5000 passes	6	12	4	8
Post-pay subscriber	8	16	6	12

This system also ensures that higher charges are made for drivers travelling during the peak periods and thus acts as a form of restraint. The tariff regime is fixed for two years and freight tariffs are double those for cars. It can be seen that, by pre-purchasing a large number of passes, discounts of up to 40% can be obtained in peak periods.

The most popular system is subscriber post-payment, whereby subscribers are billed via their bank at the end of each month according to their actual use of the system by direct debiting. Late payment incurs an excess charge of NOK 125. A family or company subscription may be obtained to include more than one vehicle in a subscription. Under the highest number of pre-paid passes, the largest discount (60% off-peak) may be obtained under this arrangement. The annual revenues projected from the system are shown in Table 17. Total annual average daily traffic is approximately 50 000 vpd; costs per transaction are around NOK 1.2.

Table 17. Projected revenues, Trondheim

	E6 NOK million	Toll ring NOK million	Total NOK million
Revenue	10	70	80
Operating cost	2	10	12

Only 1 station out of 11 on the toll ring is manned, the remainder are fully automatic. For Ranheim, on the E6, collection costs are higher since this section applies tariffs 24 hours per day.

Some of the planned expenditure under the programme for the next four years is due to be spent on the following projects:

- E6 Trondheim – Stjørdal
- E6 Nord Tangent
- E6 Nord Sør
- E6 Sluppen – Klett
- Rv 706 Omkj – vegen
- Ytre Ringveg.

Fig. 17. Paired coin collection units

Trondheim is noteworthy for its very progressive parking charge system. This acts as an additional form of traffic restraint, making extended stays more expensive by increasing the marginal cost of parking hourly, for example

- first hour: NOK 8
- second hour: NOK 18
- third hour: NOK 33.

9.5. Technology

The toll ring is similar to Oslo's with a cordon on all roads around the city comprising 11 stations. Toll collection points are either two- or four-lane, depending on the traffic volume, with automatic or manual collection systems. Only two stations are manned and none have barriers. Individual lanes are equipped as follows:

- subscription lane — overhead antenna (transmitter/receiver), automatic vehicle detector, video camera, light signal
- coin/card lane — low and high level coin collection units (Fig. 17), magnetic card reader, ticket printer, automatic vehicle detector, traffic signal, radio/speech intercom link to control centre
- attendant lane — attendant, light signal.

The automatic toll collection is based on the Q-FREE system (as in Oslo but using a different frequency). A surface acoustic wave radio frequency is transmitted to each passing vehicle and reflected from the vehicle transponder with its unique ID which is then received in the local toll station's registration system. This records the vehicle passage, time, location and lane which is transmitted and entered into the central management and accounting system for processing and customer billing.

The status of a customer's account is shown in the rapid- response signals in the automatic lane, thus:

- green light — credit
- white cross — subscription renewal due
- amber bar — not approved.

The video enforcement system records the front number plate details of the vehicle.

Over 80% of the vehicles passing at peak periods have automatic vehicle identification (AVI) tags and billing is registered automatically. The central control unit is manned and can be contacted by telephone link from each coin unit. Violators are verified visually from images recorded on the video enforcement system. The attendant toll lane is necessary at one location because of the high volume of tourists from outside town who use this road in summer.

Other options are being investigated for expansion of the charging system within Trondheim to apply, for example, to

- the introduction of electronic collection systems in parking lots
- the use of a similar pass for public transport trips, and/or
- the use of a similar pass for ferry trips.

The 'Tron' card (an integrated transport payment card or ITPC) is currently under further development and it is hoped this will result in a very progressive and fully integrated transport charging system.

In addition, expansion of the coverage or refinement into a more comprehensive urban road pricing system by zone (e.g. with three rings: inner, central and outer) and with a more complex tariff structure is under consideration. This is feasible using the existing technology with differential rates which can be applied quite simply through software modifications.

9.6. Social implications

When the first concept of introducing a comprehensive toll system was announced, it was recognized that considerable efforts would have to be made in order to convince the public of its acceptability.

The local Public Roads Administration entered into an active public information campaign whereby, through television and radio and newspaper advertising, the objectives and the benefits of the campaign were carefully spelt out. The message was focused on environmental enhancement in the city. In addition the message included the improved efficiency the system afforded to maintain a vibrant economy, and the introduction of a simple and relatively painless mechanism for collecting toll revenue.

The extent to which this programme was successful was very important. At the time of opening subscribers had been given free tags for their vehicles, which would enable cheaper electronic tollcollectionfromtheoperator'sperspectiveandgreatersimplicity (non-stop) for the user. The user would also benefit from discounts on using the system. Flexible payment alternatives were also introduced under either pre- or post-payment.

Localized boundary effects, and distortions and anomalies of the cordon were recognized. In order to overcome them, the system was modified so that any traveller could only be charged for one entry each hour. Only one area of the city was not covered by the toll ring: Trolla, to the west, where there is a low density of resident population and traffic and where, since entry of vehicles is by ferry, travellers may already be considered to have been tolled.

Generally there has not been a high level of dissatisfaction with the toll concept since implementation. It was important that evidence of the expenditure resulting from the scheme was apparent early on. The benefits from the revenue generated included the development of the ring road, Kroppen Bridge and dedicated cycle lanes in the city.

9.7. Conclusions

The Trondheim toll ring was established in October 1991. In terms of the objective of raising additional revenue for transport and environmental improvements, with a highly automated system (80% passholders), annual net revenues of approximately NOK 60 are projected to be received for the first year.

From the outset, the development into a fully operational urban road pricing system with further variations in the tariff structure and/or area of coverage have been given consideration. In the longer term it is considered that this system may provide the framework for a fully automated, integrated transport payment system which could include user charges, parking charges and public transport fares.

In terms of equity, the differential tariff structure which is higher at peak periods and which only allows free additional trips after a relatively high threshold (75 trips per month), ensures that the system is aimed at charging a price for congestion whereby the users more directly perceive the effective marginal cost of each additional trip at peak periods. The system of pre-payment enables significant discounts to be attained for use in off-peak periods. Off-peak and weekend shopping and recreational trips are not charged. The rule of charging subscribers for no more than one entry per hour through the toll ring was introduced to overcome boundary anomalies.

By maximizing the electronic charging at the tolls, the average collection cost is reduced. In addition to increasing efficiency, this automation enables more complex and variable charging structures to be introduced which are fairer and which achieve the desired distributional effects. Currently there are 24 different values of tariff applicable for differing vehicle categories and time periods.

In terms of acceptability, a thorough public information campaign was held prior to the system's introduction and incentives were given (with free tags and discounted rates) in order to encourage subscribers to pre-register with the system. This was very successful. In addition it was considered that, with increased peak period tariffs, there should be adequate periods during which travellers would not be charged in order to address the problem of congestion.

10. Stockholm

10.1. Introduction

The city of Stockholm is uniquely situated within an archipelago of 24 000 islands and inlets on the east coast of Sweden. With its history and nautical origins, Stockholm County now contains 1.6 million inhabitants, with 675 000 residing within the defined boundaries of Stockholm City (Fig. 18).

As in other Scandinavian cities, land use and transport routes are determined by the topography, historically with ferries and boats providing the key connections. Since the Second World War, with the development of the road network, traffic in the capital has grown rapidly. As a result, policy makers are urgently focusing on alternative means of addressing this issue. There is an increasing number of vehicles registered in Stockholm County, with car ownership currently standing at 366 per 1000 inhabitants.

During the 1980s, as a result of declining economic activity in urban areas, a Metropolitan Commission was established which recognized the importance of metropolitan regions to the national economy, and hence the general need to increase the level of investment in public infrastructure.

As a consequence a special National Infrastructure Fund was established with SEK 20 billion earmarked for road, rail and mass transit projects to be managed by an infrastructure investment advisory panel. Both the National Road Administration and the National Rail Administration were identified as agencies responsible for the development of specific projects. At the regional level, developments were to be dependent on joint collaboration by the government and the county and city councils.

10.2. Objectives

In 1990, recognizing the growing traffic and environmental problems, a transport plan was undertaken for Stockholm with the following goals:

- to reduce traffic-related pollution (noise and emissions)
- to reduce the number of road accidents

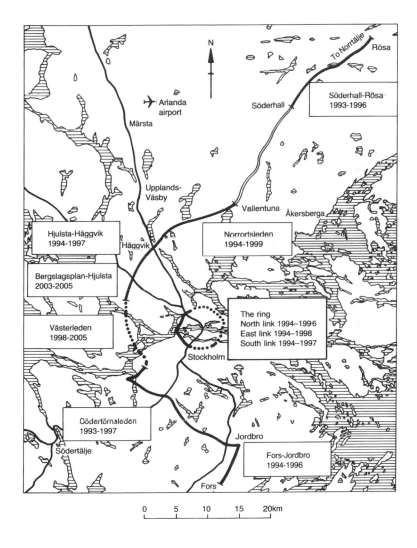

Fig. 18. Stockholm County and its planned major road developments

- to improve the traffic flow for buses
- to reduce congestion and improve the freedom of movement for goods vehicles
- to minimize congestion and improve the environment for cyclists and pedestrians.

In order to achieve these goals, an integrated solution comprising a package of measures was defined and adopted. These measures principally included

- upgrading and expanding of the public transport system
- constructing new relief roads
- introducing area tolls and stricter parking controls.

10.3. Institutional framework

In order to develop these measures in detail with the different parties concerned, the Minister of Transport and Communications in 1990 convened a special negotiator (Bengt Dennis, Governor of the Central Bank of Sweden) to undertake this process which became known as the Greater Stockholm Negotiations. This process involved delicate political negotiations with representatives of the Social Democratic, Moderate and Liberal parties in both the City and County of Stockholm. As a result, a detailed agreement was reached for a defined 15-year development programme.

The programme had an estimated investment cost initially of approximately SEK 28 billion (1990 prices), later raised to SEK 36 billion (1992 prices) for public transport and roads. To achieve this, new forms of financing would be required. It was envisaged that this should come from two levels of state and regional government allocations, and area tolls on traffic in and around the inner city. One fundamental prerequisite was that, in order to achieve the defined objectives, public transport in the region must be improved.

It was also envisaged that further aspects would have to be negotiated in order to define the complete implementation details, in particular concerning

- Österleden, the eastern inner ring road
- Västerleden, the western inner ring road
- Ytre Tvärleden, the outer cross route

Table 18. Development programme, Greater Stockholm Negotiations

Public transport	SEK million (1992)
1. Railway	
New tracks 'wasp waist' Stockholm – Arta	1550
Double tracks Kallhall – Kungsangen	1248
Double tracks Alvsjo – Västerhaninge – Nynashamn	1121
2. Subway	
Improvement, repair and renewal	6433
Hjulsta – Barkarby	290
Roslag line	611
3. Light Rail	
Ph.I Gullmarsplan – Alvsjo – Liljeholmen – Alvik	2057
Ph.II Slussen – Gullmarsplan	600
Ph.III Alvik – Bromma	1500
4. Public transport core network	
Core network – central Stockholm	350
Environmentally friendly buses	340
Express bus	230
(Financing of public transport will be by state and county authorities)	
Sub-total	15750
Roads	
1. Inner orbital ring road	10510
(southern link, eastern route and northern link)	
2. Outer cross route – Western by-pass	7650
(Fors – Jordbro, Södertorn, Västerleden, Hjulsta – Häggvik, Norrortsleden)	
(Financing of inner ring and outer cross route will be by user charges)	
Sub-total	18160
3. Street improvements	730
4. Environmental improvements	500
5. Accessibility	400
6. Park-and-ride facilities	230
Sub-total	1860
Total	35770

along with user fee charging details on these new links. The planned development programme is listed in Table 18.

The importance of overcoming the environmental disturbances caused by transport in the metropolitan area was recognized at a very high level. Because of the criticality of this issue, the need for consensus was paramount in order to attain the initiative in undertaking coherent improvements. In the past disagreement has led to paralysis which has precluded any real progress in political decision making.

Consequently, the appointment of an authoritative negotiator was important. This role was seen as instrumental in bringing the many different parties in coalition at both the county and city levels to common understanding to find a politically tenable way forward. This was particularly significant given that the programme would be based on joint financing from both state and county levels but also with emphasis on additional direct road user charging. After attainment of the final agreement in September 1992, a 'checkpoint' negotiation was set for 1996, in order to enable the targets and achievements to be reappraised at that stage.

The initial (Dennis 1) negotiations concluded that large-scale road building should be financed through road user charges. It was mandated that the fee system should be based on an electronic toll collection system and that this should

- meet defined reliability, flexibility, and legal requirements
- permit differentiation of fees with respect to time of travel, type of vehicle and type of emission control.

In addition it was proposed that the system should be phased in as follows:

- when the entire inner ring is complete (1996), a fee should be charged to travel across the ring when entering the city.
- when the entire outer western bypass is complete (1998), a fee should be charged to travel in either direction on this route.

As a result of the subsequent negotiations (Dennis 2), the supplemental agreement reached in September 1992 was determined on the following principles:

- the agreement is long-term and is based on sharing by state and county authorities of the financing requirements

- there will be significant public transport improvements, particularly in the central area, a new light rail system, and enhancements on the Mälar, Svealand and Arlanda airport lines

- the National Road Administration will construct the inner ring and outer cross routes and these will be financed by the introduction of road user charges (funds so released will enable investment to be made in public transport facilities)

- there will be environmental improvements in the central area of the city and emphasis on a core grid for bus traffic

- as a follow-up there will be continuing conferral twice a year and further 'checkpoint' negotiations in 1996.

This agreement slightly modified the previous proposals. The company which had been formed in April 1991 to implement the package (including construction of the western bypass, construction of the inner orbital ring road and implementation of a toll financing system), Stockholmsleder AB, was to be strengthened and to proceed with implementing the road development programme. State guarantees would be made available to facilitate the financing of the road construction programme.

Stockholmsleder AB is 80% owned by Väginvest (a subsidiary of the National Road Administration) and 20% by the city. It is to serve as the agent for the realization of the system. All municipalities in the county are to be offered part ownership in the company. Despite objections by environmentalists and the reservations by one political party as to the detailed design and alignment of several of the road and tunnel sections and access and ventilation arrangements, the design was to proceed with the goal of the realization in stages of the key road sections. Simultaneously, development and analysis of the toll system was to proceed with definition of the cordon, siting of the toll stations, refinement of the technology and review of the required legal and administrative arrangements.

10.4. Financial arrangements

A series of scenarios has been developed by the National Roads Administration for the introduction of road user charges. These have addressed the imposition of tolls on the proposed new outer road sections and also on a cordon imposed around the inner city itself or the possibility of an inner area charge (Table 19, Fig. 19).

Table 19. Pricing scenarios, Stockholm

		Toll: SEK		Toll: SEK
A0	Do nothing			
A1	Inner ring	10 one way	Western bypass	5 each way
			Orbital	5 each way
A2(D)	Inner ring	10 (20) one way	Western bypass	10 (20) each way
	3 central areas	5 (10) each way	Orbital	5 (10) each way
A3	3 rings	5+5+5 one way		
A4(D)	Outer ring	10(20) one way	Western bypass	5(10) each way
A5	Inner ring	10 one way		
	3 central areas	5 each way		
A6	Inner ring	10		
A7			Western bypass	5 each way
			Orbital	5 each way
A8	Inner city	0.5/min	Western bypass	5 each way
			Orbital	5 each way
A9	Inner city	0.25/min	Western bypass	5 each way
			Orbital	5 each way
A10	Inner ring	20 one way	Western bypass	10 each way
	3 central areas	10 each way	Orbital	10 each way
A11	Inner city	0.1, 0.25 or 0.75/km		
A12	Outer ring	20 one way	Western bypass	10 each way
A13	Outer ring	15 one way	Western bypass	5 each way

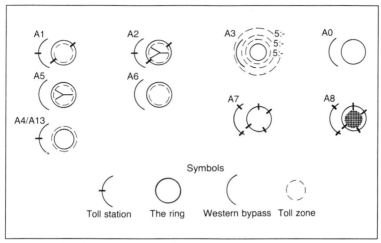

Fig. 19. Cordon Scenarios, Stockholm

In the analysis of alternatives, the following charging options were discounted as being too complex:

- time charging per minute
- distance charging per kilometre
- multiple ring cordons
- central area charges.

Orbital road charges were also considered inappropriate.

The most suitable charging system was found to be a combination based on

- an outer city cordon ring toll (approximately 25 stations)
- a western bypass road section toll (two stations).

In analysing various sensitivities a simple solution (in Table 19 option A13, similar to A4), appeared to offer the best results based on the following parameters:

- a 24 hour toll ring cordon on entering the city with maximum automation and a flat rate tariff (SEK 15/car, SEK 45/heavy vehicle) for all vehicles
- a toll point on the western bypass for all vehicles passing either way with a tariff (SEK 5/car, SEK 15/heavy vehicle) on all vehicles.

This system, estimated to cost approximately SEK 275 million to install, would achieve the desired pay-off within 20 years, reduce inner city traffic by about 30%, reduce overall journey times and enable regional traffic to pass around the city. It would also enable a central area scheme to be implemented in the future if so desired in order to enforce a more comprehensive urban road pricing system.

10.5. Conclusions

In the case of Stockholm, developments have focused on the political process in order to reach consensus on a package of measures which is acceptable to most parties and which is achievable. With the objectives of reducing congestion, improving traffic flow and public transport services, and reducing environmental pollution, the debate has focused on policy consensus and acceptability, financing methods and practical alternatives.

As a result of the supplementary negotiations, an agreement was achieved. It is too early to judge the measures from the standpoints of equity, efficiency and acceptability. The importance of the political process whereby the early achievement of a consensus, including issues of funding, can facilitate the establishment of an implementation framework in which each stage works effectively towards realizing the agreed goals, is emphasized by experience in Stockholm.

It has also been recognized that such a process should be maintained through regular follow-up negotiations along with a 'balance sheet' analysis of provisions and results.

In Stockholm, the introduction of road pricing will generate funds earmarked for road construction and improvement, so releasing state funds for allocation towards improvement of the public transport system. This clearly illustrates the need for an integrated package of measures in which toll financing provides one element of support which is essential in realizing the programme. While the structured negotiations succeeded in identifying an integrated development programme, this process needs to continue in order to win acceptance of the tariff charging policy and implementation programme.

11. Gothenburg

11.1. Introduction

Gothenburg is the second largest city in Sweden. It is a considerable industrial centre, including Volvo's manufacturing centre and a very active port on the River Gota. The old town is on the south bank of the river and the city is traversed by the E6 road, the main artery between the south including connections to Denmark and the north including routes to Norway. The city has a population of approximately 800 000, 450 000 of whom live in the central area (Fig. 20).

The city is endowed with good public transport, in particular a comprehensive network of trains and trams in the central areas. In the past the city has experimented with several parking schemes which have acted as traffic restraint measures. The main constraint on north–south communications is the River Gota.

11.2. Objectives

At the same time as the Greater Stockholm Negotiations were initiated, in 1990, a similar negotiating framework was set up in the region of Gothenburg under the chairmanship of Ulf Adelsohn. The objective was similarly to seek consensus on the establishment of a programme for the development of transport and environmental improvements in the area which will contribute to the overall economic benefits to the region.

11.3. Institutional framework

The region comprises 11 communes. One of the principal goals was to seek measures with a broad degree of support, from different constituencies, often with conflicting aspirations. Co-ordination of such planning in the region is undertaken through the Association of Local Authorities.

During early 1991 an agreement in principle was reached, which outlined a programme of transport improvements over a period of ten years. It was agreed that the process should proceed but that further discussions should be undertaken in spring 1995 to review the way in which the agreement was being implemented and to provide an effective check on the process.

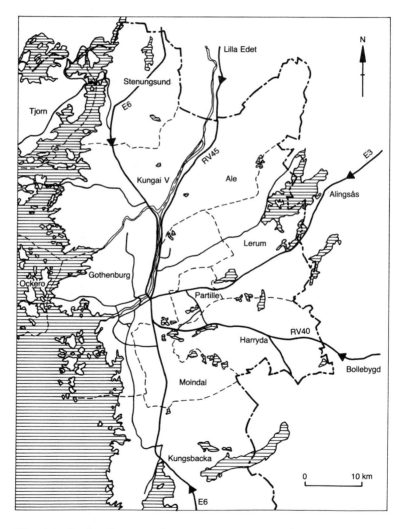

Fig. 20. Gothenburg and its major access roads

While the need for new outer road links was identified, particular emphasis was placed on avoiding the need to undertake construction in inner Gothenburg, and on environmental improvements. Such a trade-off for enhancing traffic flow in the outer areas should ensure calmer green zones in residential areas and restrict environmentally disruptive vehicles. Table 20 summarizes the principal measures outlined in the agreement.

11.4. Financial arrangements

Of the SEK 10 000 million expenditure outlined, approximately 40% was envisaged for mass transit and 60% for roads, including tunnels. The major expenditure would be on the E6 eastern bypass. It was envisaged that tolls would be levied in order to finance its construction.

With respect to other sources of finance, each of the 11 municipalities agreed to set aside SEK 500 million for investment within its own boundary. It was envisaged that the financial participation of other institutions might be considered through, for example, land

Table 20. Development programme, Gothenburg

	SEK million
Public transport	
Refurbishment/enlargement of tramways	1700
Park and ride, and bus lanes	400
West coast, Bergslagen, Boras, Bohus line; regional trains in western Sweden; and relocation of goods terminal	2230
Subtotal	4330
Roads	
Introduction of toll roads	
E6 peripheral and eastern bypass	4600
Götaleden and Brackeleden tunnels Green zone inside ring road and development of ecologically friendly vehicles	1300
Subtotal	5900
Total	10230

development revenues. In addition, regular state grants would be made available plus SEK 1300 million from the National Infrastructure Fund. However, these objectives were considered ambitious.

It was necessary to study the E6 eastern bypass in further detail. The city has a western orbital road and a bridge crossing at Alvsborgsbron but no eastern bypass. Consequently, much of the north–south traffic uses the E6 as the central artery which passes through the city. Four options were studied: the long and short rings and the long and short arteries. The latter two pass more centrally through the city, but mostly in tunnel. Each option requires a new crossing of the River Göta and some alternatives include land development over the enclosed tunnel. Traffic volumes are in the region of 50 000 vpd.

In order to finance construction, options were investigated of charging tolls of either SEK 10 or SEK 5 per trip for entry into the central area. A more comprehensive study is also being undertaken of the possibility of introducing a broader urban road pricing system throughout the city. This includes analysis of the following options through the creation of alternative charging zones, for example:

A — environmental zone (west)

B — central zone (inner)

C — developed zone (dense)

D — central and developed zone

E — peripheral ring cordon

F — cross zone

G — cross zone (five sub-zones)

H — inner city area

I — inner city area and central zone

J — greater commune area.

For a construction outturn cost of approximately SEK 7 billion (half financed by the state), the region would need to generate revenues of approximately SEK 350 million per annum over a period of 10 years.

Alternative pricing structures have been addressed ranging from toll collection just on the new road (at entry and/or exit), at a cordon of approximately 50 stations or in several zones as described above (approximately 160 stations). Traffic passing straight through or around the city would also be charged. Under the DRIVE programme, an automatic debiting system (ADS) pilot project has been tested on Dag Hammersköld Avenue.

11.5. Conclusions

The negotiations were generally successful in outlining a programme. However, the ensuing political commitment was not forthcoming because signatories insisted on ratification of the proposals by their own authorities, and this was not attained. Consequently, in the absence of local support and with no obvious central government support, this has left individual authorities manoeuvring in their own local interest. With the possibility of a plan for the reorganization of the local government framework being formulated, there is insufficient agreement on the way forward. While the agreement concluded that the process should be revisited in 1995, including review of the way that the agreement has been implemented, there is a risk that few clear commitments will have been made by then.

Compared with Stockholm, Gothenburg is wrestling with a greater number of participants in the negotiations through a looser association of local authorities and less of a consensus. Clearly, a contribution from the National Infrastructure Fund would assist with the construction of the E6 bypass but resources still remain a constraint. The municipality should be free to design its own urban road pricing system to manage the inner area traffic.

12. Malmo

12.1. Introduction

'Malmö' is Sweden's third largest city. It has a population of approximately 250 000. The main developments are centred around the port and the historical area and arterial roads radiate from the city. As an active port and with significant industrial activity, it is also considerably influenced by neighbouring Denmark. To the north lie the towns of Helsingborg in Sweden and Helsingör in Denmark, just 5 km apart. These cities have traditionally held close links and are connected by a very frequent car/rail ferry service.

The recent decision to build a fixed combined road and rail link across the Öresund, just to the south of Malmö, will also affect the transport patterns in the region in the future. With the Great Belt West Bridge and the East Bridge and Tunnel under construction to link the separate Danish islands, the Öresund crossing will complete the physical transport link between continental Europe and Scandinavia.

12.2. Objectives

The Metropolitan Cities Commission recognized the importance of the metropolitan regions to the national economy. As in the cases of Stockholm and Gothenburg, in 1990 the government established a negotiating team in order to seek solutions to improve the environment, reduce congestion and enhance development prospects in the area, which would be acceptable to all political parties. The Malmö team was chaired by Sven Hulterström.

12.3. Institutional framework

The administration in the Malmö region is complicated through having a large number of different authorities. In common with Gothenburg, the City of Malmö and the Malmohus County Council have formed a single Association of Local Authorities. Through this body, the Malmö negotiations were undertaken with the goal of outlining and agreeing a series of measures to be implemented (Table 21).

Table 21. Development programme, Malmö

	SEK million
Public transport	
Malmö – Ystad – Sturup railway	260 – 440
West coast rail line	
Southern main rail line	490
Rail tunnel in centre of Malmö,	
connecting Öresund Bridge and	
central station, and	
regional trains in southern Sweden	4000 – 5000
Roads	
Malmö outer ring road	1130
Other road improvements	874
Total	7000 – 8000

12.4. Financial arrangements

The major investment planned for the region relates to the connection with the Öresund crossing. As this connection is of such strategic national importance, the necessary finance might be obtained from the National Infrastructure Fund. The Öresund link is scheduled to be opened in June 1997. For the outer ring road (bypass) for Malmö, financing alternatives are being investigated which could include tolls.

The implementation of any of these projects is dependent on success in defining an appropriate financing structure that fits in with national funding arrangements. Several alternatives are under review. The main thrust of the programme as a whole is dependent on the development of the Öresund crossing in particular. This major project is the main factor shaping any future investment decisions.

12.5. Conclusions

It is apparent that out of the three sets of metropolitan negotiations which have been carried out, those in Malmö have been the least successful. While a set of objectives has been agreed, sufficient political consensus has not yet been achieved to enable the project to proceed. The development programme remains an outline of objectives, but the initiative has not yet been harnessed to enable any individual projects to progress.

13. The Randstad

13.1. Introduction

The Randstad constitutes a horseshoe-shaped conurbation comprising the cities of Utrecht, Hilversum, Amsterdam, Haarlem, Leiden, Den Haag, Delft, Rotterdam and Dordrecht (see Fig. 21). With a population of over 6 million people in separate centres with intervening agricultural areas, it has an impressive road, rail and waterway system. It also contains the world's largest port (Rotterdam), a major international airport (Schiphol), a significant financial centre (Amsterdam) and the centre of government (Den Haag). Haarlem, Utrecht and Den Haag are also the seats of the provincial governments of North Holland, Utrecht and South Holland respectively.

Consequently, instead of one large homogeneous metropolitan centre, there are effectively four medium-sized and other smaller polycentric cities spread throughout the region. The geography also provides constraints; the lowlands of the Netherlands are dramatically different from the fjords of Scandinavia but their network of rivers and canals still require bridging by the road or rail links.

Waterways have traditionally been the prime transport routes and they retain their importance today. Travelling passenger craft pay a fee for passage beneath any bridge that has to be raised. In spite of this, there is no general tradition of toll payment on transport routes in the Netherlands. However, with rising levels of congestion, attention has increasingly focused on this issue.

Amsterdam

The port of Amsterdam was created with the damming of the River Amstel at the Zuider Zee in the thirteenth century. After further settlement and after extensions and digging of canals, the city was walled, and it has thrived ever since. It now has 800 000 inhabitants.

Amsterdam's canals are strategically linked by the North Sea Canal to the coast and also with the Rhine Canal. As a result of housing shortages, there has been pressure on the city to establish new districts and a number have been established in the outer areas, in some cases on reclaimed polders (e.g. Schiphol on Haarlem polder and Almere and Lelystad on Flevoland).

The city has a comprehensive public transport network with metro and light rail, 16 tram lines and over 30 bus routes. These carry large numbers of passengers and are well integrated, both in terms of the network they form and the co-ordinated charge structure. In spite of this there is considerable congestion in the central area with 130 000 cars registered there.

In March 1992 a referendum was held in which all residents were asked their opinion concerning the banning of cars from the central area. In spite of a turn-out of only 28%, the City Council is proceeding with the proposal to ban most vehicles from the historic centre. Extensive measures will be required, in particular enhancements to facilitate park-and-ride, increase the length of the cycle path system and use of the canals.

An earlier infrastructure plan for the city had identified new areas of development along two axes: the River Ij axis towards Haarlem

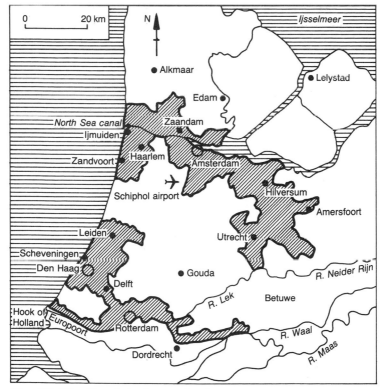

Fig. 21. The Randstad conurbation

and the southern axis through Schiphol airport. A new regional plan is being prepared which gives particular emphasis to issues of regional development.

From a transport perspective the goal is to enhance accessibility to North Holland Province and in the Randstad region overall. This is particularly important given that Amsterdam is now a triple port: with a sea port, an airport and, more recently, a telecommunications port.

In general terms the outlines of the national policy are for

- enhanced accessibility for business and freight traffic
- improved quality of life (environment) by reducing the growth of passenger car traffic.

From the city's perspective, this could be achieved through the introduction of some form of road pricing to act both as a restraint and as an additional source of revenue for improving access. The general aim is that, in order to increase general accessibility, some decrease in non-essential car mobility should be enforced within the metropolitan area. However, parking policy or simple supply constraints are currently perceived as being the best means of restraining personal commuter or non-essential travel. Emphasis is also being placed on methods to increase the capacity of existing facilities outside the inner area. For example via the following:

- advisory speeds on trunk roads such that, with a more uniform flow, higher traffic throughput volumes can be achieved
- ramp metering of access to trunk roads
- dedicated bus lanes, car pooling and higher occupancy of vehicles
- proposals to introduce dedicated freight lanes to enhance movement to/from Schiphol and Rotterdam Europoort.

In the short term particular emphasis will be maintained on these types of measures before the introduction of more politically sensitive pricing schemes for which public acceptance is far from assured. As indicated, parking policy is being used extensively in order to dissuade non-essential passenger car trips, along with ongoing subsidy of public transport operations. For example, the cross-subsidy offer of free public transport for all students with a student card has increased ridership levels and prevented their transfer to other modes.

Settlement location policy is also important, with careful consideration being given to the location of new commercial or residential communities and their relationship to the existing transport network.

Den Haag

Den Haag has a mixture of roles including the seat of state and of provincial government (South Holland), a royal residence and a diplomatic centre. It is different from most other cities in that it did not originate from commercial or trading forces. It has a population of 450 000 and is very well served by public transport.

Its transport network comprises rail, light rail and buses with dedicated bus lanes and cycle lanes in the central area. Here urban transport policy has focused on improving accessibility and environmental amenity. This is being incorporated through the enhancement of public transport services and encouragement of car parking in strategic areas such as at railway stations outside the city centre.

Particular emphasis has been given to parking policy as a restraint tool. In inner areas all street parking areas are controlled by meters or weekly and monthly permits. A number of off-street, privately-run car parks have been built.

Rotterdam

Rotterdam is part of the southern flank of the Randstad. It is located at the confluence where Europe's major rivers, the Rhine and the Meuse, meet the North Sea. Since dredging of the canal, the port has grown to substantial proportions from the original docks over the polders west of Spoorweghaven. Urbanization has also occurred on both sides of the rivers followed by satellite developments. Due to its strategic location, one-third of all goods imported and exported by the EC now pass through the Netherlands. An intricate and dense network of waterways, roads and railways has developed around Rotterdam, the hub of economic activity. The predominant mode for transporting bulk materials and petrochemicals is ship and waterway traffic with bulk, container and fuel traffic being transported through the inland canal and river routes linked to the rail network.

The rail network runs through the central urban area and connects all the docks. The road network traverses the navigable waterways by bridges and tunnels. There are major bridges to the south (Haringvliet and Moerdijk) and the north (Brienenoord), and five key tunnel crossings (Benelux and Maas – Nieuwe Maas and Botlek, Heinenoord and Drecht – Oude Maas) as well as dam crossings. Since the end of the Second World War when most of the city was destroyed, it has undergone considerable urban development including the conversion of the older dockland areas of the city to new uses.

Policy in Rotterdam has focused on the following policy issues:

- the economic position of the Rotterdam area as an economic centre and a transport hub
- the development of the main port, Rijnmond, for petrochemicals, containers and food in the western and middle zones (Europoort) and the eastern zone (Waalhaven, Eemshaven and Merwehaven)
- the development of the city's economy with office development in central locations, trade and business services at peripheral locations and transport junctions
- the provision of public service and social facilities in four central areas, and at peripheral transport interconnections
- the renovation of housing and the creation of a compact city with outer 'garden city' neighbourhood housing
- improvement of public transport, enhancement of road infrastructure and main arterial roads
- enhancement of open spaces and the quality of the environment.

In order to address these issues a series of integral tasks was defined in a regional context, namely

- to cluster facilities in the central area
- to renew pre-war neighbourhoods
- to develop the four important central areas
- to concentrate development at transport intersections
- to enhance the city-harbour zone
- to plan recreational parks
- to integrate the Ijsselmonde island and enhance the harbour area along the motorway backbone.

From a traffic perspective, given the necessity to ensure efficient freight movement in the region, the policy has been to reduce private commuter road travel through the provision of park-and-ride facilities at railway stations and car pooling points on the outer motorway network. Together with the locations of the key crossing points in and out of the central area, their potential suitability as positions for toll collection was investigated. The potential revenue capture here is substantial in terms of traffic volume, but avoiding delays and congestion is of paramount importance. For this reason there has been no consensus as to the objectives of tolling. Different groups in the region have differing interests relative to the introduction of road user charging.

13.2. Policy background

The Netherlands as a country has a tradition of fairly interventionist policies of national and regional planning. In the transport sector, the Second Transport Structure Plan[7] was designed to develop a strategy to address the fundamental problems as currently identified. By determining policies in particular areas, a series of key policy issues was defined with a view to shaping developments in the forthcoming years. Particular consideration was given to the means of facilitating movement of passenger and freight vehicles in the roads sector. These policy issues are listed in Table 22.

The establishment of an infrastructure fund (the final measure listed in Table 22) was perceived as being a means to an improved framework for operating an integrated transport policy and would combine the existing National Road Fund and Mobility Fund. Overall income would constitute some tax revenue, a proportion of motor vehicle tax and motor fuel tax, and toll income or a peak hour surcharge. One measure proposed under the plan was to increase the tax on fuel. Emphasis was also given to regulatory measures to ensure enforcement of payment.

13.3. Developments in road user charging

With the continuing growth in traffic and vehicle ownership, in the late 1980s considerable attention was devoted to researching methods of addressing the special transport needs of the urban-ized western part of the Netherlands in particular. With the review *Mobility Scenario — Randstad*[8] and with the government's initial transport proposals under the Second Transport Structure Plan (Part A),[7] the key policy objectives concerning the desired balance between mobility, individual choice and the environment were spelt out. A period of public consultation ensued and considerable modifications were made to these proposals. The following describes resultant developments with respect to road user charging.

Recognizing the potential of road pricing and that some form of demand management would be required as part of a package of measures, a project team was established under the Ministerie Van Verkeer en Waterstaat in August 1988, with the particular goal of

Table 22. Netherlands Second Transport Structure Plan, 1990[7]

Strategic goal

Sustainable development with a balance of individual freedom, accessibility and environmental amenity, achieved in stages by

- limiting external effects
- ensuring accessibility
- managing mobility

Target scenarios

By category, to achieve established targets in the following key policy areas.

Managing mobility

- Location planning — concentration of housing employment, leisure and other public facilities in relation to transport networks
- Parking norms for commercial and public facilities
- Urban remodelling — road network layout and car-free areas to discourage car use
- Application of telecommunications
- Socio-economic developments — spread of working and opening hours
- Pricing — raising variable costs of motoring, application of tolls on certain access roads, realignment of public transport fares and decreasing public transport user costs relative to that of private car trips

---

addressing the viability of road pricing, named Project Rekening Rijden (road pricing).[9]

The programme of requirements for a road pricing system was approved by the Minister in February 1989. This was geared towards achieving the stated objectives of

- counteracting congestion by making better use of the roads

- improving the environment by slowing the growth of car traffic

- raising additional finance for improving accessibility.

As a result of detailed review, the project team proposed a scheme for the region, having investigated various options of different operating systems and charging regimes, for example

Table 22 continued

Enhancing accessibility

Passengers

- Collective transport — improvement of travelling time and reliability of public transport
- Provision of cycleways
- Road network — elimination of bottlenecks and optimizing use of existing link capacity
- Encouragement of car sharing
- Information technology — travel information and traffic management systems
- Transfer facilities, e.g. park-and-ride facilities.

Freight

- Road haulage — increasing the use of the existing road capacity (vehicle loads and freight lanes)
- Dedicated rail freight lines (e.g. Betune)
- Upgrading principal waterway routes
- Combined transport — integration and container handling
- Information technology — introduction of a management information system.

Within the transport plan, a series of measures was also outlined which would need to be achieved in order to implement the goals of the policy. This included

- establishment of effective regional frameworks for administrative collaboration in the form of transport regions
- co-operation between transport organizations
- establishment of an infrastructure fund

- a network of charging stations (multiple cordon — 4 primary and 100 secondary stations)
- charging regimes of differential tariffs (NFL 2.50 peak, NFL 1.50 off peak)
- an operating system with on-board unit and a stored-value electronic transit card.

By February 1990 the functional design specification of the proposed automatic debiting system had been developed. However, it was becoming apparent that political support for the scheme was waning and that emphasis on other policy measures would be preferred. With parliament unlikely to allocate funding for the indus-

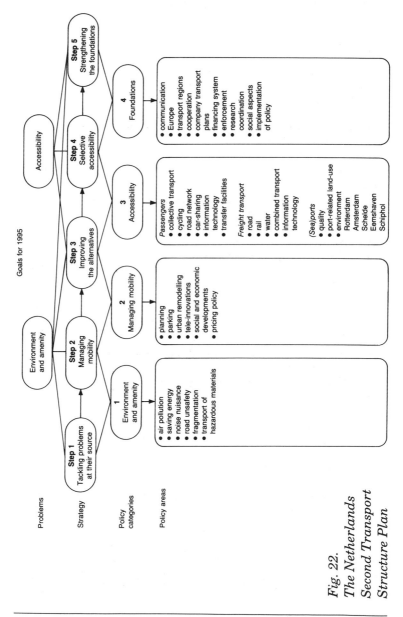

Fig. 22.
The Netherlands
Second Transport
Structure Plan

trial development of the technical equipment, the Cabinet scaled down the project and redefined the objectives.

Consequently, with the publication of the government decision on the Second Transport Structure Plan (Part D) in 1990,[7] the overall policy objectives were very clearly restated in terms of improving accessibility and overall environmental quality. A series of combined measures was targeted to reduce car commuter traffic through a policy addressing the location of employment and residential areas and a parking policy. These were considered the key instruments rather than road user charging.

With the formation of a new government in autumn 1989, the directives were re-orientated away from congestion pricing towards simpler means of collecting road user charges. The brief of the project team was revised and it was renamed Project Tollheffing (toll collection) in 1991. The revised aim was to develop a simple system of toll collection on the national trunk road network in the Randstad region. This was perceived principally as a method of generating additional funds for financing the transport policy and the construction of five new tunnels which formed particular bottlenecks in the region (Noord, Wijker, Benelux, Coen and Blankenburg), and to achieve some measure of restraint of traffic volumes in congested areas. This was a considerable variation from the objectives as initially outlined, moving from an essentially urban to an outer-urban framework.

While the project was being developed, resistance again began to develop, this time with respect to the unacceptability of large and numerous toll plazas throughout the area. Other systems (such as 'vignettes'), were also considered by another research group to restrain travellers at peak periods. After a further review, such mechanisms as peak-hour surcharges to the annual vehicle licence were considered administratively too complex and were abandoned as unviable.

More recently, the policy objectives have again been redefined so as to address the problems of congestion in the area. With the economic dependence on freight movements and the role of the region as a trans-shipment centre and hub of trans-continental freight movements, the goal has been revised to restrict the mobility of non-essential movement by using the pricing mechanism to influence driver behaviour. This would ensure that trips would be either transferred to other time periods, suppressed or eventually transferred on to other modes.

Of the three alternatives of urban road pricing, trunk-road toll collection or peak charging, the latter has now become the focus of the strategy for implementation. Consequently, after a further revision to the project team's brief, in June 1992 it was restructured and renamed Project Spitsbydrage (peak contribution) in order to address methods of charging traffic at peak periods only. The goal is to dissuade non-essential vehicles from competing for limited road space at peak periods. This remains the target of the ongoing research.

13.4. Conclusions

Developments in the Randstad are very interesting from a public policy perspective. Over the past five years government has actively sought viable solutions to overcome increasing traffic congestion. In the context of the Randstad, with a series of polycentric nuclei, the issue is to maintain good access between centres rather than, as in other cities, within urban centres. Parking policy is generally used as a restraint within the cities of Amsterdam, Rotterdam and Den Haag. Pricing policy has been regarded as a mechanism to be applied to trips generally outside and between centres.

The objectives as currently defined have evolved from those established earlier. From an initial emphasis on urban road pricing (through cordons around urban areas) to improve accessibility and environmental quality and to generate additional funds for new infrastructure, they were modified to direct toll collection to raise revenue. Now the focus is aimed at reducing congestion at peak periods through the levying of charges at peak times.

These changes have occurred principally due to two factors: the established objectives themselves had certain intrinsically contradicting goals; and public opinion served to redirect the emphasis towards environmental and quality-of-life objectives.

In terms of acceptability there is little enthusiasm on the part of the public towards the idea of a pricing policy. However, as the problem of peak period congestion is recognized, if developed in a reasonable framework, a combined policy through pricing and other equitable measures to defer or dissuade non-essential trips at peak periods and/or to encourage diversion to other modes may be found acceptable. Clearly this has to be orientated towards making car drivers aware of the real costs of their travel and being charged accordingly, and increasingly passing on freight transport costs to the user.

The motivation in the Netherlands, with its economic dependence on its position as the hub of European freight transport is to maintain an efficient transport network. This will require each traveller to be charged to pay for the real cost of each trip he or she makes.

14. London

14.1. Introduction

In the UK, traffic congestion is most readily apparent in almost any area of London. In 1992, with a population of 2.5 million in Inner London and 7 million in Greater London, car ownership stood at 370 cars per 1000 people. Average traffic speeds in the capital are only 11 mph, and as people struggle to avoid peak hour congestion the peak period is lengthening. In addition it is no longer just the inner areas that are congested, but the outer areas and boroughs too. Traffic congestion now affects most parts of London for seven hours per day and, in the central parts, for nearly twelve. It has been estimated that the cost of this congestion in and around London is £10–15 billion per annum.

14.2. Background

The introduction of road pricing has been on the agenda in the UK for a long time. In the report of the Ministry of Transport's *Investigations into the economic and technical possibilities of road pricing* in 1963,[10] the Smeed Committee concluded that, with the deficiencies of the existing methods of taxation, the introduction of direct user charges would be required to address the economic effects of traffic congestion.

The emphasis then was on how various forms of taxes could be levied to differentiate between congested and uncongested roads, and the benefits that each form would produce. While the full potential and social issues were not addressed, it appears that the Committee had assumed that any direct user charging introduced would be accompanied by a reduction in existing taxes so that the total motoring population would pay no more than before. Of the suggested methods it was considered that, while parking tax and a system of daily licences were advantageous, a direct road pricing system via metering, if it could be developed in the long run into a viable system, would yield considerable benefits.

This report was followed by one entitled *The better use of town roads* in 1967[11] which — on recognizing that technical difficulties would take time to resolve and in seeking more urgent methods to restrain traffic — reverted to emphasis on comprehensive parking control. However, parking control does not adequately address means of optimizing use of the overall road space.

At about the same time, Buchanan's influential report *Traffic in towns*[12] was published and it stressed that with growth in both the number and use of private vehicles increasing, investment would also need to increase. If the quality of life in towns was to be maintained at a civilized level, the report asserted, there was a limit to the amount of traffic that could be accommodated. With the establishment of environmental standards, this factor would determine the accessibility of a town. However, accessibility could be increased, but only in accordance with the amount of money that was spent on physical alterations to the road network and the road hierarchy.

14.3. Institutional framework

The London County Council (LCC) was succeeded by the Greater London Council (GLC) with responsibility for overall strategic planning and transport in the urbanized area. Due to its differing political stance with the national government and the resultant tensions, it was abolished in 1986 after 21 years in existence, and its responsibilities were devolved to the 32 individual London boroughs.

There were two particular anomalies with the GLC: contradictions resulting from the fact that it was originally established as a highway authority rather than as a transport authority, and the incompleteness of its geographical coverage (ideally the built-up area bounded by London's green belt). The M25 motorway now provides a physical boundary, although this cuts through the once protected green belt in several areas (Fig. 23). Discussion continues as to the best method of representing the interests of the inhabitants of London. Many consider that a 'slimline' elected authority would be preferable to the current situation.

Despite the creation of a small planning group called the London Planning Advisory Council (LPAC), there is now no overall strategic planning authority for the city. This prevents anything more than a piecemeal and incremental approach being taken to transport sector investments. In 1992 a Traffic Director for London was appointed. This new role was to focus on the development of a network of priority (red) routes across London aimed specifically, through enhancement of traffic controls and traffic management, at reducing congestion and journey times on key axial routes. The role was to co-ordinate the traffic management measures of the different highway authorities in relation to these priority routes with the aim of achieving

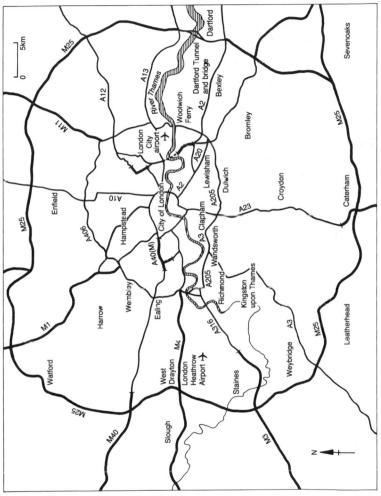

Fig. 23. The major road routes in the area bounded by London's orbital motorway, the M25

- reduced journey times
- improved reliability of bus journey times
- reduced accidents
- improvements for pedestrians and cyclists.

A key element of this process relates to enforcement of kerbside controls and to the management of the often competing demands for road space of different users.

14.4. Research framework

With subsequent developments in road transport and information technology and the particular experiments in Hong Kong in 1983–1985 (see chapter 17), in 1991 the renamed Department of Transport initiated a £3 million research programme in the region bounded by the M25. This was to cover all aspects of differential pricing and in particular to address

- demand — travel choice by user group
- supply — transport network and public transport systems
- technology — functional requirements
- the institutional and administrative framework
- the financial framework.

In addition, the research was to focus on the potential impacts of the introduction of an urban road pricing system, particularly with respect to

- urban economy
- environment and safety
- equity
- public acceptability.

An assessment framework was established in order to undertake a coherent appraisal of the potential costs and benefits of alternative schemes. This will enable the individual impacts to be assessed and their particular effect on different behaviour groups to be determined. Consequently, the impacts on each behaviour group (by spatial, socio-economic, business, public service, transport operator, journey purpose and mobility categories) will be evaluated (see Tables 9 and 10).

Some of the initial research into public reactions has indicated that there is considerable support for the introduction of road pricing if it is to be combined with other elements of traffic restraint and enhancement of public transport. Particular requirements cited are the need to

- improve public transport coverage, co-ordination and service levels
- improve road traffic management
- enhance regulation and enforce parking controls
- introduce traffic calming, environmental enhancements and restrict 'rat running'.

The financial benefits need to be applied to these areas. Particular issues relate to the need to suppress non-essential trips, to encourage increasing the number of riders per vehicle and to maximize use of the available road space. The research programme is focusing on modelling and behavioural issues in order to enable anticipated attitudes to be addressed and the suitability of emerging technology and electronic systems.

With respect to the technology, comprehensive road pricing within the confines of the M25 would necessitate an extensive zonal and cordon system. The scale of the operation would mean a large number of vehicles would have to be equipped and daily transactions dealt with, and many roadside beacons would need to be emplaced. In order to implement a fully automatic system, an effective enforcement system would be required. As technology develops and increasingly intelligent transponders and smart cards become available, it may become feasible for road users to maintain their own 'electronic purses' of credit to pay for road use, parking and public transport.

14.5. Existing toll systems

There are currently two toll systems in operation in London which are significantly different in scale and nature: the Dartford River Crossing and the Dulwich College Estate Tollgate.

Dartford River Crossing

The Dartford River Crossing toll was established in 1991 with the opening of the privately-financed cable bridge in parallel with two older tunnels under the River Thames. This comprises a key link in the M25, where peak traffic flows exceed 100 000 vpd. Tolls per crossing are £0.80 for cars, £1.50 for LGVs and £2.30 for HGVs at any time of day.

An automatic debiting system has been introduced using a smart card called a DART-tag. Subscribers pay a minimum initial payment (cars £50, LGVs £75, HGVs £125) for a personalized tag to be attached to their vehicle's windscreen. On passing the toll plaza a signal electronically identifies the tag, activates the barrier and deducts the cost from the card. The card is kept in credit through the pre-purchase of units by the user on a monthly basis. When the card is running short of credit, signals at the toll plaza indicate this with an amber light. Subscribers to this system receive a 7.5% discount compared to the cash toll and their transaction and waiting time at the toll booth is reduced.

Dulwich Tollgate

The Dulwich Tollgate has been in existence since 1789. In return for permitting the construction of a road and providing access across Dulwich College's private grounds, a charge has been levied for over 200 years. With the building of the Crystal Palace nearby in 1854, traffic grew significantly on this short section of road, making it a valuable turnpike.

Currently, traffic volume averages approximately 1000 vpd one way. The method of charging changed little over the years until full automation in 1993. Charges are as follows: motorcycles pass free of charge and cars incur £0.50, payable using a coin or card acceptor. A 250 unit stored value decrementing card gives the owner a 15% discount per passage. Charge periods are 0700–2130 hours weekdays, 0800–1800 hours Saturdays and 0900–1700 hours Sundays. HGVs are banned.

14.6. Conclusions

In spite of the continual debate over deteriorating traffic conditions in the London area, little has been achieved on the supply side to alleviate conditions. The Downs–Thomson paradox remains — that is, if road capacity is increased under congested conditions and draws travellers off the public transport system, the attractiveness of the road system deteriorates. Conversely, by improving the public transport system, the road conditions may be improved. Any solution to this complex problem requires a package of physical, financial and regulatory measures, focused particularly on the demand side.

Deteriorating traffic conditions are evident in many other cities in the UK besides London. For this reason, metropolitan authorities are increasingly looking more seriously at local road user charging solutions. The Department of Transport's initiative in undertaking a detailed research programme, while welcome, needs to be translated into action in order to define a positive framework upon which can be developed effective urban road pricing systems that afford a tangible benefit to the maximum number of people.

14.7. Ongoing debate – inter-urban and urban roads

In the UK roads sector a debate is currently being waged at several different levels:

- How should roads be provided – by the public or private sectors?
- How should they be financed – by direct or indirect charges?
- How can traffic congestion be overcome – by pricing, restraint or demand management?

At one level, stimulated by the Department of Transport's publication "Paying for better motorways", the debate is focused on motorways and inter-urban roads:

- whether private sector finance can be harnessed to fund the development programme;
- whether private companies should take over ownership of parts of the road network;
- whether charging should be introduced to facilitate improvements to the highway network;
- whether direct road user charging should be introduced either electronically or via a permit; or
- whether the benefits (of reduced congestion and easier journeys) would outweigh the costs and potential diversion.

In addition, the debate runs deeply as to the potential for market style solutions, to increase the productivity of the highway system and facilitate its development in response to the demands of its users.This includes appraisal of the appropriate ownership structure of the highways and the whole issue of privatisation, assessment of the benefits of unencumbered ownership, franchises/concessions or management contracts. Generally, the increased involvement of the private sector can lead to depoliticization, focus on financial investment criteria, generation of transparency of costs and speedier response. But to enable this, equity has to be created from property rights in the highway system, through clarification of the ownership of the roads - without this, electronic road pricing will be hybrid and not a market activity. This would provide the framework for charging for road-track costs.

As one commentator has suggested - overall, the best method to solve congestion would be to have the roads provided by the private sector and vehicles by the public sector!

From the general public's perspective, the overriding concern is that any user charges should be hypothecated and retained in the transport sector. Charging should not be introduced solely as a method of earning additional revenue for the Exchequer. Suggested figures of 1.5p per car mile, or 4.5p per HGV mile, which are estimated to generate up to £1 billion per annum, are, however, seductive to the Treasury.

The mechanics of phasing in a system of direct user charging on primary roads on a national basis are being addressed. A transition from indirect to predominantly direct charging could be instituted via an electronic "vignette", with a gradual evolution from card tax discs to electronic tags or in-vehicle units. Other alternative ways to support private sector development through the introduction of "shadow tolls" are also being explored. Already under EC law, heavy goods vehicles over 12 tonnes have to purchase tax discs to use the motorways in the Benelux countries, in Denmark and in Germany.

Meanwhile, at another level, in urban areas, the emphasis is on continuing research to assess possible benefits and disbenefits of introducing direct user charging via a system of either area licencing, cordon charging or peak period pricing to reduce congestion. A variety of scenarios have been tested for London – as shown in Figure 24. A number of detailed surveys are continuing to be undertaken to ascertain the likely and varied reactions of different users.

14.8. Other metropolitan areas – Cambridge, Edinburgh and Bristol

In addition to the Department of Transport's ongoing research programme into congestion charging in London, other cities are pursuing their own agendas. In the region of Lothian (Edinburgh) and Avon (Bristol) in particular, different scenarios have been modelled to assess the possible benefits and impacts of the introduction of alternative forms of electronic road pricing. Hampshire and other counties are developing a transport strategy to reduce reliance on the private car and to enhance other modes of transport. This will require a balanced policy of restraint, regulation and pricing.

Demonstration programmes have been commenced. In Cambridge, part financed by EC DRIVE under the ADEPT programme,

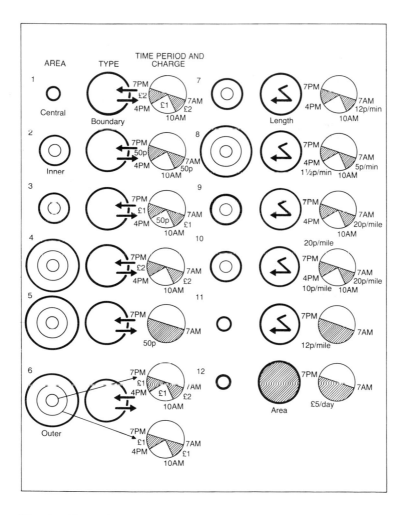

Fig. 24. Tested scenarios, London

a small scale pilot scheme has been initiated, investigating the feasibility of in-vehicle charging for the time spent travelling particular distances through areas demarcated by beacons. This congestion metering system levies a charge when a vehicle has passed a specified time and distance threshold (currently proposed as 3 minutes to travel 0.5 km). A similar type of demonstration scheme charging by time and zone rather than distance in the London Borough of Richmond has not yet progressed.

Direct charging experience to date in the UK has been restricted to toll facilities on new bridges and tunnels, or on established facilities, predominantly estuarial crossings, for example:

- Bridges – Severn, Forth, Humber, Dartford, Skye, Tay, Irskine, Tamar, Itchen, Clifton, Whitchurch, Cleddau, Shrewsbury, etc..

- Tunnels – Mersey, Medway, etc..

The first privately financed toll roads (Birmingham Northern Relief & Western Orbital Road) are still in the planning stage where substantial issues (e.g. planning permission, toll locations, traffic diversion) still remain to be addressed.

15. France

15.1. Introduction — inter-urban roads

Whereas in Scandinavia, Norway has a tradition of tolls predominantly from the need to pay for ferry crossings to enable access across rivers or fjords, in France the development of the national motorway network has been encouraged through the granting of concessions by the State to individual companies to develop and operate specific sections of autoroute across France. Whilst the US national interstate network was developed primarily by defence motives, the French network has been developed to enhance access between centres and to promote the "désenclavement" of particular regions.

Whilst still centralised individually around Paris, as shown in Figure 25, key links have been constructed in other Départements. As a consequence many areas are now served both by the regular Routes Nationales, which are non-tolled, and also by new sections of autoroute which are tolled and which provide a premium level of service. Charges range between 0.3 FFR/car kilometre and 0.6 FFR/HGV kilometre. Consequently, for inter-urban travel, the option for the general user is either a direct but generally less well-designed tolled route, or a less direct, untolled and generally slower route.

Between 1956 and 1963, several concession companies (Sociétiés d'Economie Mixtes) were established to develop sections of toll road. Essentially public companies, they were under-capitalised and required subsidy. During the 1970s, they were restructured, but financial difficulties continued. In 1982, three out of four companies were nationalised and the government harmonised toll rates with cross subsidies to support the financially weaker concession companies. Today there is one wholly private company, Cofiroute (740km), created in 1970, in the Loire region, and six others in different regions:

- SANEF(du Nord et de l'Est de la France) – 1400km
- AREA(Rhône-Alpes) – 440km
- SAPN(Paris-Normandie) – 480km
- ESCOTA(Esterel-Côte d'Azur) – 430k
- ASF(du Sud de la France) – 1640km

- SAPRR(Paris-Rhin-Rhône) – 1350km.

Some further amalgamation is still anticipated to form probably four fully regional autoroute companies with coherent networks. Total overall annual income of all of these companies currently averages approximately FFR 15 billion. Several automatic collection systems are in use on the autoroutes – e.g. Télépéage, Tolltag, Premid, etc..

In addition to inter-urban autoroutes, there are several road facilities in France which are tolled (péages). These are generally bridges or tunnels as in other countries, where charges are made due to the monopoly nature of the facility, for example:

- Bridges – Tancarville, Normandie, Aquitaine, etc.

- Tunnels – Strasbourg, Sainte-Marie aux Mines, Mont-Blanc, Fréjus, Marseilles Prado-Carenage, etc.

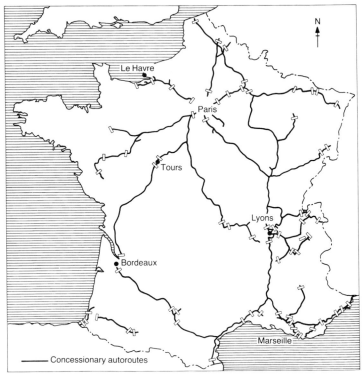

Fig. 25. Autoroutes, France

Whilst generally of coin collection, increasingly the use of smart cards is being adopted for toll collection for subscribers, but few truly non-stop systems have yet to be introduced. A harmonised automatic debiting system for HGV vehicles is being developed jointly by different toll station operators.

15.2. Metropolitan areas – Paris, Marseilles and Lyons

In contrast to the extensive experience of tolling on inter-urban roads in France, there has been less experimentation to date with road user charging in urban areas. Charging has been focused on parking. Three schemes are, however, of interest, in the major metropolitan areas of Paris, Marseilles and Lyons.

- Paris – on the A1 autoroute into Paris, differential charges are now being introduced at peak periods predominantly at week-ends as car drivers incur congestion on returning to Paris.

- Marseilles – at the Prado–Carenage Tunnel, a central cross harbour two level car tunnel, drivers are charged variable rates.

- Lyons – on the inner urban ring road being developed with private and public finance, tolls will be charged on entering the network.

Other major developments under consideration for Paris include underground highway tunnels (e.g. Réseau MUSE and Rocade Souterraine Péripherique) in the Ile de France, which will require tolls to finance them.

16. Singapore

16.1. Introduction

Singapore is an island state with a population of approximately 2.7 million. Under a fairly strict regime the country has industrialized very rapidly and constitutes a very dynamic economy. During the 1970s and 1980s the GNP per capita grew at rates averaging 7% per annum. The car ownership rate is 180 per 1000 inhabitants.

16.2. Objectives

With a large population, a vibrant economy and only very limited physical resources, the demands on Singapore's road network are excessive, particularly at peak periods. For this reason, in 1975 a traffic restraint scheme was designed with the objective of restraining traffic flows at peak periods into or through the central business district to alleviate congestion. With emphasis on a series of traffic management measures, the goal was to release pressure on the road

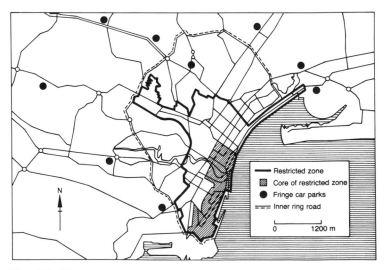

Fig. 26. Singapore

network and provide alternatives, and to improve the flow of buses without restraining commercial traffic by ensuring that only licensed or high occupancy vehicles would be able to travel into the restricted central area.

16.3. Institutional framework

With only one level of government the Public Works Department (PWD) had a direct mandate to implement an effective and functional traffic restraint scheme. In order to implement the licensing scheme the following complementary traffic management methods were introduced:

- an increase in parking charges (for both private and public parking places) and higher rates on more central locations
- park-and-ride facilities just outside the restricted area
- shuttle buses between outer parking areas and the central business district
- encouragement of flexible working hours
- a peak-hour ban on commercial vehicles with more than two axles entering the central area.

At the same time a new mass rapid transit system was under development and a bypass, the East Coast Parkway, was being constructed. The government has a firm policy of restricting vehicle ownership in this small country by means of taxation. On a new vehicle there is

- 45% import duty on a vehicle's cost
- a basic registration tax of S$1000 for private cars and S$5000 for company cars
- an additional registration tax of 150% of a vehicle's cost.

The latter may be modified by a preferential annual registration fee if an old vehicle is scrapped and replaced by a (less polluting) new vehicle.

Once registered, an annual road tax applies which is proportionate to engine size. A special category for registering cars for leisure use was introduced in 1991, with lower registration tax and annual tax. These 'weekend cars' have different red number plates and can only be used off-peak and at weekends.

In 1990 the government introduced a quota system whereby certificates of entitlement are required in order to buy a new vehicle.

Each month the number of vehicles allowed to be registered is determined and purchasers bid for the right to import and register a new vehicle. In consequence, the cost of car ownership is very high.

16.4. Operational framework

The scheme which was adopted was an area licensing scheme and on its introduction in 1975 it comprised

- a restricted zone of over 5 km^2 in the central business district (see Fig. 26)
- restricted periods 0730 –0930 hours Mondays to Saturdays
- restrictions on private vehicles except taxis, buses, cars with more than four passengers and commercial vehicles.

In order to enter into the restricted zone drivers have to display a licence on their vehicle windscreen (Fig. 27). The licences are dated and identified by shape (a daily is rectangular and a monthly circular) and a colour code which varies monthly. These can be acquired at kiosks on a daily or monthly basis and are checked visually by wardens stationed at 33 control gantry positions. This forms in effect a cordon around the restricted zone (725 ha) as vehicles pass into the city without stopping. Any violators are liable to a fine that is imposed by mail.

Shortly after its introduction, the exemptions on taxis were withdrawn. In addition, the rates for company cars (with 'Q' registration plates) were doubled over those for private cars. In 1989 the following revisions were made:

- restricted modes of operation were increased to include afternoon peaks from 1630 –1900 hours Mondays to Fridays
- the list of restricted vehicles was extended to include buses, pool cars, commercial vehicles and motorcycles
- tariffs for all vehicles were reduced from S$5 to S$3 per day, or $100 to $60 per month.

16.5. Financial framework

The area licence fees which allow multiple entry and their evolution are as shown in Table 22. In the early years unit rates were increased from S$3 to S$4, then to S$5 per vehicle before being reduced back to S$3. Rates have not changed since 1990.

While the goal is primarily to act as a restraint to traffic, net monthly revenues of approximately S$0.5 million were obtained

Fig. 27. Singapore area licence

Table 23. Licence fees, Singapore

	1990 Whole day * S$	1990 Part day † S$
Motorcycle	1	0.7
Private car	3	2
Company car	6	4
Taxi, bus	3	2

* 0730–1830 Mon.–Fri. and 0730–1500 Sat.
† 1015–1630 Mon.–Fri. and 1015–1500 Sat.

initially. In 1992 annual revenues were S$38 million from license sales. With traffic volumes of approximately 45 000 vpd in the morning peak and 25 000 vpd in the afternoon peak, transaction costs are very low with only about 40 attendants undertaking enforcement. The initial capital cost (gantries and signing) was low. The monthly operating cost is approximately S$0.3 million. In this respect, this 'low-tech' scheme is both effective and low-cost leading to significant time savings from reduced congestion.

16.6. Social impacts

After the introduction of the scheme, traffic fell dramatically in the morning peak, but travellers tended to reschedule their trips to just before or after the restriction. This resulted in the authority extending the morning charge period to 1015 hours. On increasing the cost of parking, public transport became more attractive, although neither the park-and-ride scheme nor the special shuttle buses was ever very popular. With a penalty of S$50, the level of violation remained under 1%.

The impact on the afternoon peak was less successful since only inbound traffic was charged. While cross-town traffic was reduced, traffic diverted and increased significantly on the ring road. The afternoon charging period was subsequently reduced to end at 1830 hours and to apply between 1200 and 1400 hours on Saturdays to accommodate business interests. A further reduction of the morning charge led to a net increase in morning peak traffic again, but charging did lead to a reduction of commercial traffic in the peak periods.

16.7. Conclusions

The scheme initially achieved traffic reduction levels at peak periods of up to 40% — higher than the goal of 25–30% — along with increased traffic speeds. The current plan is to replace the existing manual/visual scheme with a system of automatic electronic debiting. Three trial schemes have been established for experimentation part funded by government contribution. These demonstration projects at TUAS will be based on electronic road pricing (ERP) and will comprise in-vehicle units with smart cards. With the current very high cost of vehicle ownership, this would enable policy to be varied to relate more directly to vehicle usage.

In terms of efficiency the area licensing scheme has low capital investment costs and physical requirements, low operating costs, and is simple to implement and enforce. It is flexible yet unsophisticated and has proved to be a successful and economical method of traffic restraint.

From the perspective of equity, the scheme has been effective, partly due to a series of related traffic management methods that has enhanced public transport. The use of higher occupancy vehicles and the suppression of non-essential trips have been encouraged. The opening of the excellent MRT light railway system in 1990 has helped to transfer ridership to the public transport system.

From the viewpoint of public acceptability, the scheme has proved to be generally popular as a means of preventing further increases in congestion. The very high cost of vehicle ownership is becoming generally prohibitive and in itself acting as a restraint on private cars. The results of the electronic trials are awaited with interest.

17. Hong Kong

17.1 Introduction

Economically Hong Kong has been one of the most dynamic of the South-East Asian countries, averaging continuous rates of growth in GDP per capita of approximately 5–10% per year. With a population of around 6 million living at a very high density, car ownership in the territory is only about 70 vehicles per 1000 inhabitants but it has the highest number of vehicles (200) per kilometre of road space in the world. With the doubling of real income over a period of ten years, the number of registered vehicles also doubled, the majority of these being private cars. Despite the uncertainty relating to the transfer of the territory to the People's Republic of China in 1997, the country, including Hong Kong Island, Kowloon peninsula and the New Territories, remains economically extremely dynamic. The tradition of tolled facilities is also well established, with charging on the Cross Harbour, Lion Rock and Aberdeen tunnels.

17.2. Policy framework

In 1976 a comprehensive transport study was undertaken to address the measures required to overcome increasing traffic congestion. As a result the government determined in its internal transport policy to

- improve the road infrastructure
- expand and improve the mass transit system
- make better economic use of road space.

In order to achieve these aims, in 1982 the policy adopted was to focus on drastic fiscal restraint to contain traffic. Particular measures introduced were similar to those adopted in Singapore, namely

- to treble the annual licence fees for private cars
- to double the first registration taxes of private cars to between 70 and 90% of the vehicle import price
- to double duty on petrol.

17.3. Institutional framework

In Hong Kong, which is a British dependent territory until 1997, the head of the government is the Governor. The Governor is assisted by an Executive Council and a Legislative Council. While the Executive Council is appointed by the Governor, half of the Legislative Council is elected. The highest policy-making group within the government is the Government Secretariat, which includes the Transport Branch. With the approaching transfer of Hong Kong in 1997 to become a special autonomous region of the People's Republic of China, a series of 19 district boards were established with the goal of increasing representation in the outer districts. Public consultation exercises were held with the districts in order to determine their reactions to the proposed fiscal and road pricing policies, but these were purely advisory.

17.4. Objectives

Due to the high cost of ownership and the fiscal regime imposed, private vehicle ownership fell over a period of several years (from 211 000 in 1981 to 170 000 in 1984). However, private car and taxi use still remained high, particularly during peak periods. Parking policy had clearly reached the bounds of practibility as a restraint. In order to address this issue it was decided by the Transport Branch of the Government Secretariat to undertake an experiment to implement direct road user charging on private vehicles using a prototype of electronic road pricing. This pilot project (costing HK$ 40 million) was undertaken between 1983 and 1985. The goal was to assess the viability of introducing a comprehensive road user charging system using electronic road pricing (ERP) on a larger scale.

17.5. Technology

The pilot project was based on a system of automatic vehicle identification (AVI) with a passive electronic number plate (ENP) the size of a video cassette, mounted on the underside of the vehicle. At a number of toll locations, a series of induction power and receiver loops was embedded in the road pavement, linked to roadside computers. Each passing of a vehicle was detected, the vehicle interrogated and identified and the data transmitted via dedicated lines to the control centre. Here, a management and accounting system could process the data and bill motorists automatically on a monthly basis. Closed-circuit television was used to detect violators by photographing the rear number plates of vehicles.

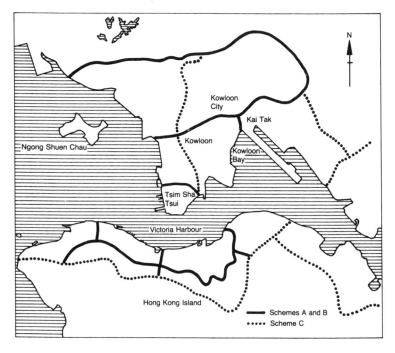

Fig. 28. The cordon schemes tested in Hong Kong

The engineering system was run for a period of 8–12 months in the Central/Admiralty area. In all, 2600 vehicles — half of which were government-owned and one-quarter of which were buses — were fitted with electronic number plates and 18 on-street sites were equipped. The control centre was equipped to handle approximately 30 000 transactions per day.

17.6. Financial framework

Three different cordon schemes were tested with different charging characteristics. The schemes were each based on a multiple cordon system, applying different charges in each area. Schemes A and B were designed to restrain radial movement, while scheme C (with an additional north–south cordon) was designed to address orbital movements. Each area is bounded by cordons and has its own distinct charge. Location of the cordons was determined by Hong Kong's geography and this resulted in an egg-timer shaped arrangement, with Kowloon and the New Territories to the north; the Tsim Sha Tsui central business district and the harbour crossing as the neck, and Hong Kong Island to the south. This arrangement, based on a series of differently priced cordons rather than on a single cordon, has more impact on longer distance commuters. The main characteristics were as follows:

- Scheme A
  - Area — five zones and 130 toll locations.
  - Period — shoulder period 0730–0800 hours, morning peak 0800–0930 hours, inter-peak period 0930–1700 hours, evening peak 1700–1900 hours, and shoulder period 1900–1930 hours.
  - Tariff — peak HK$ 2, HK$4 or HK$6 per cordon; shoulder HK$1, HK$2 or HK$3 per cordon; tidal surcharge HK$1 per cordon, off-peak (1900–0730 hours Monday–Friday, all day Saturday and Sunday) no charge

- Scheme B
  - Area — five zones, 115 toll locations, no boundary tails
  - Period — as scheme A
  - Tariff — as scheme A plus directional tidal surcharge of HK$1

- Scheme C
  - Area — 13 zones, 185 toll locations
  - Period — as scheme A
  - Tariff — as scheme A (all commercial vehicles were excluded).

As an example, a north–south private car trip from Tsuen Wan to Tsim Sha Tsui under scheme A would cost as follows according to the time of travel:

- shoulder period: HK$ 3+2+1 = HK$ 6 (zones 1, 2 and 3).
- peak period: HK$ 6+4+2 +1 = HK$13 (zones 1, 2 and 3) including surcharge
- inter-peak: HK$ 3+2+1+1+1 = HK$ 8 (zones 1, 2 and 3) including surcharges.

The estimated costs of installing the system were HK$ 240 million and the estimated annual operating costs were HK$ 20 million per annum. The estimated revenue was forecast to be HK$ 400–500 million per annum.

The average driver would therefore have to pay about HK$ 120 per month, but this would be offset by the government's intention to lower annual vehicle licence fees by HK$ 100 per month.

17.7. Social implications

During the pilot project a series of surveys was undertaken. Based on a travel characteristics survey, a set of impact models was developed to predict the changes in travel demand resulting from the introduction of electronic road pricing due essentially to anticipated changes in travel mode, trip time or destination. These were compared with alternative methods of introducing either an area licensing scheme, a car ownership restraint scheme or optimal road pricing charges. The results indicated that the introduction of electronic road pricing would produce significantly greater benefits to the community than an area licensing scheme. This is essentially due to the flexibility of the former's charging regime, and the fact that it offers the car user a choice as to how to react to the charges.

17.8. Conclusions

Despite many encouraging indications as to the potential benefits of the scheme and the viability of the technology, it was decided not to proceed beyond the pilot project stage of the ERP system. There were several reasons for this.

- The stock and property market crashes in 1982 had led to a decline in real income.

- The fiscal restraint measures were still being felt in 1985.

- Private car owners felt severely discriminated against and considered ERP as being punitive.

- Political uncertainty remained with the signing of the Sino-British declaration in 1984.

- The charging mechanisms of the system were considered an invasion of privacy.

- The public were sceptical that the introduction of ERP would be accompanied by a reduction in annual licence fees and first registration taxes.

- Construction of the MTR light railway system was to lead to improved public transport services.

- There were many reservations as to whether the system would actually lead to a reduction in traffic congestion.

In terms of efficiency, the net benefits of the scheme were estimated to be substantial, and as a policy tool it was evident that the introduction of multiple cordons would enable the charging structure to be more easily adapted to suit particular requirements.

From the viewpoint of equity, metering was presented as being the fairer option as it is based on actual use rather than on ownership restrictions — this was the theme presented under the government slogan, 'A fairer way to go'.

However, the acceptability of ERP proved to be problematic. After swingeing increases in vehicle taxes in 1982, ERP was always perceived by the public as just another mechanism to raise taxes. In addition, despite the government having proved the viability of the technology, the monitoring, reporting and billing of individual trips was seen as an unacceptable invasion of privacy. Further public consultation exercises revealed significant opposition to the scheme.

A second comprehensive transport study was undertaken and in 1990, following a period of consultation, revised options (renamed 'area pricing', but extensively similar to the previous Scheme B extended to include Saturdays) were proposed by the government. With the political uncertainties relating to the transfer in 1997, it is unlikely that any comprehensive road pricing system will be introduced. As in other cities, a package of measures is considered to be necessary with the revenue to be collected by a special authority and earmarked for road capacity and public transport improvements.

In conclusion, it was considered, as a result of the pilot project, that the road pricing system was not a panacea but that it could act as a useful component in a balanced transport policy once road improvement, public transport and traffic management initiatives had been exhausted.

18. USA

18.1. Introduction – inter-urban highways

The first turnpikes had originated in Europe, where individual owners charged for the use of their section of road or bridge ("pontage"). Users suffered from the ever demanding owners in conflict with their presumption of a right of passage, such that by the end of the 19th Century most turnpike trusts had been disbanded.

In the USA, over 10 000 miles of roads have been developed by turnpike companies. Similar facilities still persist with the development of tolled sections of highways and expansion in individual States in the inter- and post-war years, particularly on the East Coast, with the Pennsylvania (1940), Maine (1947), New Hampshire (1950) and New Jersey (1951) turnpikes being financed by revenue bonds.

Other key urban roads are the Connecticut Turnpike; New York Thruway; Garden State Parkway; Ohio Turnpike; Indiana Tollroad; Illinois Tollway; Florida Turnpike and Kansas Turnpike. Typical toll rates are in the range of US $0.05–0.10 per km.

At a national level, the 1944 Federal Aid Highway Act called for the designation of a national system of Inter-State and Defence Highways. Its development was to be driven by defence objectives to achieve a comprehensive highway system between States across the continent, including urban highway by-passes (see Fig. 29). In 1956, a Highway Trust Fund was created whereby federal user fees were deposited in a specific fund earmarked for road improvements. Whilst State Highway Agencies are responsible for managing federally developed roads, these national federal expressways and the Interstate system (I roads) were developed to operate on a toll-free basis.

More recently, specific facilities (toll bridges and tunnels) have been developed by individual States on the eastern seaboard which generally levy charges. While remaining public authorities, these agencies are generally state-owned and are responsible for the development, financing and operation of their own facilities. In such a diverse country, facilities differ, but examples of typical tolled facilities and public operators are:

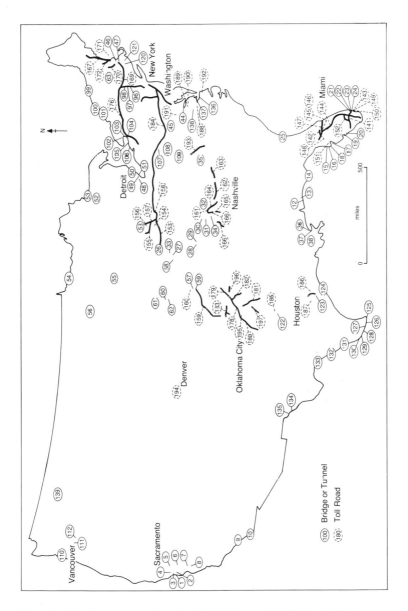

Fig. 29. Bridge, tunnel and toll highway facilities, USA

- Chesapeake Bay Bridge and Tunnel District
- Port Authority of New York and New Jersey bridges and tunnels
- Delaware River Port Authority bridges
- Golden Gate Bridge, Highway and Transportation District, California
- Triborough Bridge and Tunnel Authority.

In 1991, the US Federal Government passed the Intermodal Surface Transportation Efficiency Act (ISTEA). This, through focusing on enhancing the roads sector, opened up funding for toll roads beyond that previously permitted on federal aid facilities (roads, bridges and tunnels). For the first time, private entities could own toll facilities. In California, this was facilitated by the passage of Bill AB680 and similar laws in Arizona encouraged the private funding of the roadway system. This has resulted in the creation of several new organisations, constituted with the sole task of developing and realising new facilities, financed on a private sector basis. These are

- Dulles Toll Road Corporation, Virginia
- California private Transportation Company, Irvine, California (SR91)
- San Joaquin Hills Transportation Corridor Agency, Costa Mesa, California.

In 1992, the Department of Transportation initiated the Intelligent Vehicle – Highway System (IVHS) research programme. This is a six-year US$600 million programme aimed at the creation of a fully automated highway and vehicle system test track by 1997. The programme is broadly based focusing on the following activities:

- ATMS (Advanced Traffic Management Systems)
- ATIS (Advanced Travel Information Systems)
- AVCS (Advanced Vehicle Control Systems)
- CVO (Commercial Vehicle Operations)
- APTS (Advanced Public Transportation Systems)
- ARTS (Advanced Rural Transportation Systems).

Under ATMS, the key areas are Automatic Debiting Systems (ADS) and non-stop Electronic Toll Collection (ETC). With the preponderance of coin collection at tolled facilities and the increasing volume of traffic passing through them, emphasis has moved to the development of dynamic charging facilities. Particularly interesting examples are now in operation at

- Tappan Zee Bridge, Spring Valley, New York State Thruway Authority (E/Z pass)
- Oklahoma Turnpike, Tulsa and Oklahoma City (Pikepass)
- SR91, Orange County, California (Intellitag)
- Crescent City Connection Bridge, New Orleans, Louisiana (Tolltag)
- Georgia 400 Extension, Atlanta (Cruise Card)
- Dallas North Tollway, Texas (Tolltag)
- Chicago North–South tollway (I-pass).

18.2. Metropolitan areas

All of the above experiments are related to electronic toll collection on highways, bridges, tunnels or turnpike facilities. The emphasis now is on pursuing research on congestion pricing or peak period differential pricing for particular facilities. There is no specific experience to date on area-wide pricing in urban or metropolitan areas, but a series of five demonstration projects have been initiated under the congestion pricing research programme. These include the San Francisco–Oakland Bay Bridge peak pricing scheme; differential charging in the Orange County SR 91 highway; charging single occupancy vehicles on the HOV lane on the San Diego I-15 highway and urban congestion alleviation on I-95 (Maryland and Virginia).

19. Summary of Part 2

From current evidence around the world, a comprehensive auto-mated urban road pricing system has not yet been implemented fully. However, it can be seen that considerable experience has been gained from the schemes introduced or tested to date, particularly

- Bergen's manual cordon
- Oslo's AVI and cordon
- Trondheim's AVI and cordon
- Area licensing in Singapore
- Hong Kong's area/multiple cordon.

Their key characteristics are summarized in Table 24.

In summary, the systems reviewed have the following relative advantages:

- the Singapore scheme is the most cost-effective with the lowest capital and operating costs
- the Oslo scheme generates the highest volume of revenue
- in terms of marginal cost pricing, the Trondheim scheme most closely charges road users for the external congestion costs they impose
- the Singapore scheme acts as probably the most effective restraint.

The introduction of electronic road user charging technology, which is being gradually introduced on inter-urban roads can enable considerable flexibility in achieving particular objectives at an urban level.

As outlined in Part 1, the design of an effective urban road pricing system by an authority should be shaped according to the particular requirements of the community and implemented in concert with both supportive policies and physical measures. A cordon or a hybrid method incorporating both direct charging and/or area licensing may be appropriate. The net benefits should be returned and applied directly to enhance transport and the local environment in the interests of the local community in question.

Table 24. Comparative characteristics of road user charging systems

	Bergen	Oslo	Trondheim	Singapore	Hong Kong
Population: millions	0.3	1.0	0.25	2.6	6.0
Operating system	Manual	AVI	AVI	Area licence	Multiple cordon
Charging method	Manual	Manual/ ETC	Manual/ ETC	ALS	ERP
Capital cost	Low	High	High	Low	High
Operating cost	Low	High	Med/high	V. low	Low
Transaction cost	NOK0.5	NOK1.3	NOK1.2	S$ 0.15	HK$ 0.5
Revenue : cost	6:1	4:1	3:1	10:1	2.5:1
Tariff structure	Fixed	Fixed	Differ- ential	Fixed	Variable
Tariff duration: hours/day	16	24	11	4	12
No. of tariff periods	1	1	2	2	5
Payment method	PAYG	PAYG pre/post	PAYG pre/post	Pre	Pre/post
Periods (peak/off peak)	Uniform	Uniform	Differing	Uniform	Differing
Automation	Manual	Pass transponder		Sticker	ENP
Exemptions	Bus	Bus, m'cycle	M'cycle	Bus	HGV
Supporting measures	Parking restraint	Discount	Env. improved	Park-and- ride	Parking restraint
Disbenefits	Labour intensive		Labour intensive		Privacy

References

1. DOWNS A. *Stuck in traffic: coping with peak hour traffic congestion.* 1992, Brooking Institution & Lincoln Institute of Land Policy, Washigton, D. C.

2. ZETTEL R. and CARLL R. *The basic theory of efficiency tolls: the tolled, the tolled-off and the untolled.* 1964, US Transport Research Board, No. 47, Washington, D. C.

3. JONES P. *Review of available evidence on public reactions to road pricing.* 1992, Report of the UK Department of Transport. Her Majesty's Stationery Office, London.

4. GOODWIN P. *The 'rule of three'. A possible solution to the political problem of competing objectives for road pricing.* 1989, Transport Studies Unit, Oxford.

5. MINISTERIE VAN VERKEER IN WATERSTAAT. *Second Transport Structure Plan.* 1990, Rijkswaterstaat, Den Haag.

6. NORWAY STORTING. *Public Roads Act No. 27,* 1989, Norway Storting, Oslo.

7. *The Netherlands Second Transport Structure Plan.* 1990. Government of the Netherlands, Den Haag.

8. RIJKSWATERSTAAT. *Mobility scenario – Randstad.* 1989, Rijkswaterstaat, Den Haag.

9. MINISTERIE VAN VERKEER IN WATERSTAAT. *Rekening Rijden — road pricing in the Netherlands.* 1989, Rijkswaterstaat, Den Haag.

10. MINISTRY OF TRANSPORT *Road pricing – the economic and technical possibilities.* Report of the committee chaired by Smeed. 1964, Her Majesty's Stationery Office, London.

11. MINISTRY OF TRANSPORT. *The better use of town roads.* 1967, Her Majesty's Stationery Office, London.

12. MINISTRY OF TRANSPORT. *Traffic in towns.* Report of the committee chaired by Buchanan. 1963, Her Majesty's Stationery Office, London.

13. DEPARTMENT OF TRANSPORT. *A review of technology for road use pricing in London*. Report on London congestion pricing. 1993, Her Majesty's Stationery Office, London.

14. HAU T. *Congestion charging mechanisms: an evaluation of current practice*. 1992, World Bank, Washington, D. C.

15. HARRAP P. *Charging for road use worldwide – an appraisal of road pricing, tolls and parking.* 1993, Financial Times, London.

16. DEPARTMENT OF TRANSPORT. *Paying for better motorways*. 1993, Her Majesty's Stationery Office, London.

17. CATLING I. (Ed.). *Advanced technology for Road Transport*. IVHS and ATT, Boston, 1994.

Bibliography

1. Congestion, traffic restraint and demand management

DOWNS A. *Stuck in traffic: coping with peak hour traffic congestion*. 1992, The Brookings Institution & Lincoln Institute of Land Policy, Washington, D. C.

MOGRIDGE M. *Travel in towns. Jam yesterday, jam today and jam tomorrow?* 1990, Macmillan, London.

WARD C. *Freedom to go – after the motor age*. 1991, Freedom Press, London.

2. Congestion pricing

CONFEDERATIONOFBRITISHINDUSTRIES. *Transport in London – the capital at risk*. 1989, CBI, London.

DUPUIT J. On the measurement of the utility of public works. *Annales des Ponts et Chaussees* (1844), in *International Economic Papers 2*, 2nd series, vol. 8, 1952, pp. 83–110.

EVANS A. Road congestion pricing: when is it a good policy? *Journal of Transport Economics and Policy*, 1992, **26,** Sept., 213–243.

HAU T. *Economic fundamentals of road pricing — a diagrammatic analysis*. 1991, World Bank, Washington, D. C.

HILLS P. *Automated variable pricing for the use of road space – an idea whose time has arrived*, 1990, Transport Studies Unit, Oxford.

MORRISON S. A survey of road pricing. *Transportation Research* 1986, **20A,** March, No. 2, 87–97.

PIGOU A. *Wealth and welfare*. 1920, Macmillan, London.

VICKERY W. Congestion theory and transport investment. *American Economic Review*, 1969.

WALTERS A. *The economics of road user charges.* 1968, World Bank, Washington, D. C.

WHITELEGG J. (Ed.). *Traffic congestion – is there a way out?* 1992, Transport Geography Study Group, Leading Edge, North Yorkshire.

3. Road user charging and traffic restraint

CIOT. *Paying for progress.* 1992, Chartered Institute of Transport, London.

FRIENDS OF THE EARTH. *Less traffic, better towns.* 1992, Friends of the Earth Trust, London.

GOODWIN P. and JONES P. *Road pricing – the political and strategic possibilities.* 1989, European Conference of Ministers of Transport Round Table 80, Paris.

ICE INFRASTRUCTURE POLICY GROUP. *Congestion.* 1989, Thomas Telford, London.

MAY A. Traffic restraint, a review of the alternatives. *Transportation Research*, 1986, **20A**, March, No. 2, 109–121.

RICHARDS B. *Transport in cities.* 1990, Architecture, Design & Technology Press, London.

MINISTRY OF TRANSPORT. *Road pricing – the economic and technical possibilities.* Report of the committee chaired by R. Smeed, 1964, Her Majesty's Stationery Office, London.

4. Urban road pricing systems

DEPARTMENT OF TRANSPORT. *A review of technology for road use pricing in London.* Report on London congestion pricing. 1993, Her Majesty's Stationery Office, London.

DRIVE. Research and technology in advanced road transport telematics. 1992, DRIVE, Brussels.

HEWITT P. A cleaner, faster London: road pricing, transport policy and the environment. 1989, Institute for Public Policy Research, London.

OECD. *Intelligent Vehicle–Highway Systems — a review of field trials.* 1992, OECD, Paris.

KAWASHIMA H. *Informatics in road transport.* 1991, Keio University, Tokyo.

5. Implementation

HAU T. *Congestion charging mechanisms: an evaluation of current practice.* 1992, World Bank, Washington, D. C.

7. Oslo

A/S FJELLINJEN Annual reports 1990 and 1991, Oslo.

DIRECTORATE OF PUBLIC ROADS ADMINISTRATION. *Toll road system in Oslo.* 1990, PRA, Oslo.

LARSEN O. and HJARTHOL R. *The toll ring in Oslo – reports on travel behaviour.* 1991, Institute of Transport Economics, Oslo.

LARSEN O. and RAMIDEH F. *Road pricing as a means of financing investments in transport infrastructure – the case of Oslo.* 1991, Institute of Transport Economics, Oslo.

SOLHEIM T. *Effects of the toll ring in Oslo.* 1990, Institute of Transport Economics, Oslo.

WAERSTED K. Automatic toll ring – no stop electronic payment systems in Norway – systems layout and full-scale experiences. 1992, *Proc. Conf. on Road traffic monitoring and control.* Institution of Electrical Engineers, London.

8. Bergen

BERGEN KOMMUNE. Transport plan. 1991, Bergen Kommune.

BRO-OG TUNNELSELSKAPET. Annual reports 1990 and 1991.

LARSEN O. and RAMIDEH F. *The toll ring in Bergen, Norway – the first year of operation.* 1988, Institute of Transport Economics, Oslo.

9. Trondheim

HOVEN T. The Trondheim toll ring – the political process and the technical solutions. 1992, *Proc. Conf. on Road traffic monitoring and control.* Institution of Electrical Engineers, London.

MICRO DESIGN. *Q-FREE Systems.* 1991, Micro Design AS, Trondheim.

PUBLIC ROADS ADMINISTRATION. *The automatic toll ring in Trondheim*. 1991, PRA, Trondheim.

TRØNDELAG BOMVEISELSKAP A/S. Annual report, 1992, Trondheim.

TRØNDELAG TOLL ROAD COMPANY. *The Trondheim toll ring – facts about the toll ring*. 1992, Trondelag Bomveiselskap A/S, Trondheim.

10. Stockholm

DENNIS B. *Dennis 1 – The Greater Stockholm Negotiations on traffic and environment*. 1991, Government of Sweden, Stockholm.

DENNIS B. *Dennis 2 – Agreement on enhancing the traffic infrastructure in the Stockholm region*. 1992, Government of Sweden, Stockholm.

GELLSTEDT B. *Road user charging*. 1992, Vägverket, Stockholm.

INGO S. *Renewal of transport infrastructure*. 1992, Regional Planning Office, Stockholm.

MALMSTEN B. *Traffic and environment in the Stockholm region*. 1992, Stockholm Council.

MINISTRY OF TRANSPORT AND COMMUNICATIONS. *Traffic for growth and a better environment*. 1991, Government of Sweden, Stockholm.

STOCKHOLMS GATUKONTOR. *Trafikanalyser infor Trafikplan '92*. 1992, Stockholms Gatukontor.

STOCKHOLMSLEDER AB. *Building to improve the Stockholm environment*. 1992, Stockholmsleder AB.

TEGNER G. *Road pricing in Stockholm – some model simulated experiences*. 1991, OECD/CETUR, Paris

11. Gothenburg

ADELSOHN U. *Gothenburg agreement*. 1991, Government of Sweden, Stockholm.

GOTHENBURG REGION. *Principstudier kring vägavgiftssystem*. 1992, Gothenburg Region.

12. Malmo

HULTERSTRÖM S. *Malmo agreement*. 1991, Government of Sweden, Stockholm.

13. The Randstad

BOYCE B., LEBLANC L. and JANSEN G. *Optimal locations of toll stations in a road pricing system – an investigation into the feasibility of optimization methods*. 1991, TNO Institute of Spatial Organization, Delft.

MINISTERIE VAN VERKEER IN WATERSTAAT. Programme of require-ments. 1988, Ministerie van Verkeer in Waterstaat, Den Haag.

MINISTERIE VAN VERKEER IN WATERSTAAT. *Rekening Rijde*n. 1989, Ministerie van Verkeer in Waterstaat, Den Haag.

POL H. *The investigation of the Dutch Rekening Rijden system*. 1990, Rijkswaterstaat, Den Haag.

STOELHORST H. and ZANDBERGEN A. The development of a road pricing system in the Netherlands. *Traffic Engineering & Control*, 1990, Feb., 66–71.

14. London

BLYTHE P. T. and HILLS P. J. Adept Project – 1: Overview. *Traffic Engineering and Control,* Feb. 1994.

CLARK D. J. *et al.* Adept Project – 3: Congestion metering, the Cambridge trial. *Traffic Engineering and Control,* April 1994.

DARTFORD RIVER CROSSING LTD. Annual report, 1991, London.

DEPARTMENT OF TRANSPORT. *Research programme on road pric-ing for London*. Progress reports 1992–1993, Her Majesty's Stationery Office, London.

HALL P. *London 2001*. 1989, Unwin Hyman, London.

HEDIN J. and SUDBEREY J. Adept Project – 2: Field trials of auto-matic debiting in Sweden. *Traffic Engineering and Control,* March 1994.

HIBBS J. and ROTH G. *Tomorrow's way – managing roads in a free society*. Adam Smith Institute, 1992, London.

HURDLE D. *Road pricing for London*. 1990, London Boroughs Association, London.

JONES P. *Selling the concept of road pricing to the public.* 1993, IBC Technical Services Conference, London.

MOGRIDGE M. *Travel in towns.* 1990, Macmillan, London.

MUSTAFA M.A.S. Adept Project – 4: Multi lane electronic tolling and mono lane video enforcement systems – Thessaloniki. *Traffic Engineering and Control*, May 1994.

MVA CONSULTANCY *Transportation strategic advice – scenario testing exercise.* 1990, London Planning Advisory Council.

NATIONAL ECONOMIC DEVELOPMENT OFFICE. *Amber alert – relieving urban traffic congestion.* 1991, NEDO, London.

SHELDON R., SCOTT M., and JONES P. Exploratory social research among London residents. 1993, PTRC Annual Meeting.

TURNER D. The priority (red) route network. Proc. Conf. on Traffic congestion. 1992, British Standards Institute, London.

15. France

UNION DES SOCIÉTÉS FRANÇAIS D'AUTOROUTES À PÉAGE — Annual Report, 1993.

16. Singapore

MENON A. and SEDDON P. *Traffic in the central area – volume characteristics.* 1991, Singapore Public Works Department.

PUBLIC WORKS DEPARTMENT . *Area Licensing Scheme, Singapore.* 1991, Singapore Public Works Department.

WATSON P. and HOLLAND E. *Relieving traffic congestion, the Singapore Area Licensing Scheme.* 1978, World Bank working paper No. 281, Washington, D. C.

MENON A P G ET AL. Singapore's Road Pricing System: its past, present and future, *ITE Journal,* Dec. 1993, Washington, D.C.

17. Hong Kong

CATLING I. and HARBORD B. Electronic road pricing in Hong Kong: 2. The technology. *Traffic Engineering & Control*, 1985, Dec., 608–615

DAWSON J. Electronic road pricing in Hong Kong: 4. Conclusion. *Traffic Engineering & Control*, 1986, Feb., 79–83.

DAWSON J. and BROWN F. Electronic road pricing in Hong Kong: 1. A fair way to go? *Traffic Engineering & Control*, 1985, Nov., 522–529.

GOVERNMENT OF HONG KONG. *A fair way to go*. 1985, Government of Hong Kong.

GOVERNMENT OF HONG KONG. *Results brief*. 1985, Government of Hong Kong.

HARRISON B. Electronic road pricing in Hong Kong: 3. Estimating and evaluating the effects. *Traffic Engineering & Control*, 1986, Jan., 13–18.

HAU T. Electronic road pricing – developments in Hong Kong 1983–89. *Journal of Transport Economics & Policy*, 1990, May, 203–214.

18. USA

INTERNATIONAL BRIDGE, TUNNEL AND TURNPIKE ASSOCIATION. *Toll Financing - proud heritage, bright future*. 1991, Washington, D.C.

IVHS AMERICA. *Intelligent Vehicle–Highway Systems in the USA*. 1992, IVHS America, Washington, D. C.